BROOMRAPES of the WORLD

BROOMRAPES *of the* WORLD

Chris Thorogood
Renata Piwowarczyk
Fred Rumsey
Gianniantonio Domina

Kew Publishing
Royal Botanic Gardens, Kew

Royal Botanic Gardens Kew

First published in 2026 by
Royal Botanic Gardens, Kew, Richmond, Surrey, TW9 3AB, UK. www.kew.org

ISBN 978-1-84246-851-7
eISBN 978-1-84246-852-4

British Library Cataloguing in Publication Data
A catalogue record for this book is available from the British Library.

Design: Nicola Thompson, Culver Design
Production Manager: Georgie Hills
Proofreading: Sharon Whitehead
Index: Sharon Whitehead

Printed in Great Britain by Short Run Press

FSC www.fsc.org MIX Paper | Supporting responsible forestry FSC® C014540

EU Authorised Representative: Easy Access System Europe Oü,16879218. Mustamäe tee 50, 10621, Tallinn, Estonia (email:gpsr.requests@easproject.com).

For information or to purchase all Kew titles please visit shop.kew.org/books or email publishing@kew.org.

Kew's mission is to understand and protect plants and fungi, for the wellbeing of people and the future of all life on Earth.

Kew receives approximately one third of its funding from Government through the Department for Environment, Food and Rural Affairs (Defra). All other funding needed to support Kew's vital work comes from members, foundations, donors and commercial activities, including book sales.

CONTENTS

ACKNOWLEDGEMENTS

We are deeply grateful to the members of the University of Oxford Botanic Garden Danby Patrons' Group for sponsoring this publication, in particular, Sarah Taylor, Margaret and John Leighfield, Helen Potts and an anonymous donor.

Our sincere thanks go to Lauren Baker who supported and coordinated the significant undertaking of drafting selected taxonomic descriptions with a team of dedicated biology students at the University of Oxford. To these students we are deeply grateful: Agata Czernecka, Angharad Thomas-De Paul, Cassie Lee, Daisy Yiangou, Diana Gusta, Gareth Johnson, Harveet Purewal, Heidi Wheatley, Jessica Feng, Jiya Soomal, Joanne Forster-Martin, Julia Morgan, Laura Flenley, Mairi Franklin, Marcus Ashworth, Nancy Locke, Reuben Nebbett-Blades, Serena Johnson, Tongtong Lin, Weina Jin and Xinye Zhang.

The authors would like to thank Hélène Tronc for cultivating the *Orobanche* collection at the University of Oxford Botanic Garden, which was a valuable living resource during the preparation of this book.

Finally, we are thankful to the many botanists, naturalists and enthusiasts around the world who were so generous in allowing us to reproduce their photos – it is a far richer book for them. In particular we thank: Abdenour Kheloufi, Abdul-lateef Ismail, Adriano Soldano, Akihiro Nishimura, Aleksandr Ebel, Alexander Fateryga, Alfredo Reyes Betancort, Ambady Sasi, Andrew Hankey, Anoop P. Balan, Antonio Pujadas Salvà, Athol Ferguson, Attila Mátis, Avinoam Danin, Ben Jones, Besnik Fetiu, Bob Harwood, Charalambos S. Christodoulou (Flora of Cyprus), Christodoulos Makris (Flora of Cyprus), Corinne Merry, Craig Ontl, Daba G. Chimitov, Daniel Pavon, Dar Ben-Natan, David Schmidt, Donald R. Tomberlin, Doug Von Gausig, Dylan M. Neubauer, Ehoarn Bidault (Missouri Botanical Garden), Enrique Triano, Eric Powell, Erin Miller, Francesco Santoro, Francisco Serrano Ortiz, František Lamla, Gerald D. Carr, Gigi Laidler, Gonzalo Moreno Moral, Grahame Preston, Groeneweide Farm, Ismail Ebrahim, Jani Järvi, Joey Bom, Joey Santore, José-Vicente Ferrández, Justyna Kasińska, Karen L. Barry, Kenji Suetsugu, Majed Aldughayman, Marko Doboš, Mario Mairal Pisa, Mark Johnston, Martin Purdy, Mayur Nandikar, Mayur Prag, Michael Heine, Michael Pollich, Miguel Sequeira, Mimoun Bachiri, Mirco Gruppl, Muangpai Suetrong, Mu-Lin Chang, Neal Kramer, Nicola van Berkel, Oleg Kosterin, Paul Hankamp, Pere Fraga, Petr Kosachev, Prajwal J. Ullal, Ramazan Murtazaliev, Rebecca Wheeler, Ridha El Mokni, Ron Vanderhoff, Sebastian Hatt, Sibel Bostancıoğlu, Siddarth Machado, Srinivasan Kasinathan, Stefan Mayr, Stephen Mifsud, Sue Marsden, Ted Smit, Thomas Croze, Toby Nowlan, Todd Boland, Tony DeSantis, Ümit Subaşı, Uwe Grabner, Vinolperera, Vladimir Kolbintsev, Yael Orgad, Youp van den Heuvel, Zbigniew Iwański and Zhi-Xiang.

Hyobanche sanguinea
South Africa (left) photo: Mayur Prag

INTRODUCTION

Broomrapes (family Orobanchaceae) are among the most curiously beautiful and least understood plants. These plants are parasitic on, and attach to the roots of, other plants to extract their nutrients. The family has a wide global distribution, and broomrapes occur in diverse habitats on a wide range of host plants. Many broomrapes are entirely devoid of functional leaves, roots and chlorophyll (holoparasitic) and possess a remarkable, rather ghostly appearance – rising out of the ground, seemingly from nowhere.

In this book we adopt a broad definition of a broomrape (a name traditionally used for the Old-World genus *Orobanche*) and cover all holoparasitic genera in the family. The name broomrape derives from broom and the Latin *rapum* for 'turnip' – in other words, a turnip of broom. This refers to the swollen stem base of *O. rapum-genistae* attached to the roots of its host (typically broom). The name is now used to describe all species in the *Orobanche* genus and in its broadest sense, other genera in the family, for example the Indian broomrapes (genus *Aeginetia*), and clustered broomrapes and desert broomrapes (genus *Aphyllon*).

In recent years, broomrapes have sparked interest among plant enthusiasts in a similar way to orchids. Like orchids, broomrapes are renowned for being taxonomically challenging. Plants that appear distinct in the field are prone to becoming blackish-brown and featureless when pressed, which has led to the incorrect determination of herbarium specimens in many of the genera. Confusion has been compounded by intra-specific variation in some species (for example in the genera *Orobanche* and *Cistanche*), confusing synonymies, incorrectly cited nomenclature, and species with poor type specimens and descriptions. Nevertheless, identification of broomrapes in the field can be straightforward with careful examination. This book seeks to give clear and detailed accounts of a range of species from all the main genera around the world, so that they may be identified reliably.

Photographs and illustrations have been selected to capture as much of the diversity across genera as possible in this book. Many of the photographs were kindly provided by other botanists and enthusiasts so that the family could be well represented geographically. Ultimately this book seeks to promote awareness and dispel confusion about the broomrapes – a family of striking and fascinating botanical enigmas.

Cistanche deserticola
Inner Mongolia, China (left)

Parasitic plants of the world

Parasitic plants contain some of the most bizarre and fascinating organisms in the plant kingdom and they have intrigued scientists for centuries. They include ecosystem engineers that enhance species richness, economically significant pests on cultivated crops, and some of the most bizarre and fascinating examples of morphological reduction and reproductive biology in the plant kingdom. The largest of all flowers, *Rafflesia*, is a parasitic plant, which lives for most of its life cycle within the tissues of its host, a tropical vine (*Tetrastigma*).

Despite their intrigue, many of the ca. 4,750 or so accepted species (1.6% of all angiosperms) and 292 genera that evolved across 26 families (Nickrent, 2020) are poorly known to science: most in fact have never been cultivated; many have been seen by few – recorded or collected once, never to be observed again.

Parasitic plants, by definition, are reliant on other photosynthetic plants, to varying degrees, for their nutrition. Green plants produce their carbohydrates by photosynthesis – they are known as photoautotrophic (or carbon-autotrophic). By contrast, plants that receive some or all of their carbohydrates from other organisms are called heterotrophic. Heterotrophic plants can be separated, by the nature of their hosts, as being either mycoheterotrophic (parasitic on fungi) or haustorial parasites (parasitic on other plants). The latter penetrate the vascular tissues of their hosts using specialised organs called haustoria. The broomrape family is the most diverse group of haustorial parasites, with representatives in every biome.

All parasitic plants attach to, and feed from, the roots and stems of other plants, but their life cycles vary. Parasitic plants are diverse in form, ranging from herbaceous plants to trees, and a handful of highly reduced parasites that grow embedded in their host and lack all trace of leaves and roots (endoparasites) – *Rafflesia*, the world's largest flower, for example.

Haustorial parasites evolved twelve times independently from free-living plants, almost exclusively among the flowering plants (angiosperms). A single heterotrophic (parasitic) non-flowering plant – a gymnosperm – is known to science: *Parasitaxus usta* (Podocarpaceae). The evolution of parasites on different hosts has been accompanied by a reduction in genome size (especially the size of the plastid genome), as well as horizontal gene transfer (HGT) – acquiring new genes from their hosts through haustoria, which may help a parasite to synchronise its physiology to that of its host's, avoid host's defences and tune in to its metabolites (Yang et al. 2016; Xu et al. 2022). Parasitic plants show an intriguing pattern of convergence as a consequence of their shared biology; indeed, unrelated mycoheterotrophic and haustorial species possess a remarkable superficial similarity that can cause confusion.

Rafflesia arnoldi
Sumatra (left)

Hemiparasites and holoparasites

Haustorial parasitic plants have been classified in various ways over the centuries on the basis of their physiological, anatomical or morphological features. The two most consistently defined groups are (i) those that possess chlorophyll and photosynthesise, but obtain some or all water and minerals from their host (hemiparasites; either facultative or obligate); and (ii) those that lack chlorophyll and photosynthesis, and instead derive all water, minerals and photo-assimilates from their hosts (holoparasites). In short, hemiparasites possess photosynthetic leaves with chlorophyll; holoparasites lack functional leaves and chlorophyll. Among all parasitic plants, over 400 species are holoparasites, but the rest, above 90%, are hemiparasites. Independent from the distinction made between hemiparasites and holoparasites, parasitic plants can be classified further by their morphology and anatomy, for example, by the part of the host plant they penetrate with their haustorium: root parasites attach to the roots of their hosts (broomrapes included), while stem parasites (e.g. dodders and many mistletoes) attach to their hosts' aerial parts; endoparasites (including *Rafflesia*) live entirely within the tissue of their host plants.

In reality there exists a spectrum between hemiparasites and holoparasites (sometimes even within a single genus), but in this book we follow this classification in broad terms, for simplicity and consistency with other work. The broomrape family is notable for including a diversity of both hemi- and holoparasites. It would be beyond the scope of a single volume to cover both (the diversity of hemiparasites in the Orobanchaceae is vast), so here we cover exclusively the holoparasitic species: in other words, any plant in the family Orobanchaceae that lacks chlorophyll.

Harveya capensis South Africa (below), photo: Mayur Prag
Rhinanthus a hemiparasite, Georgia (right)
Orobanche foetida Portugal (page 14)
Lathraea rhodopea Macedonia, Greece (Page 15)

Broomrapes of the world: taxonomy and diversity

Such unusual and striking plants have unsurprisingly attracted attention for millennia, including in ancient Greece. Theophrastus was familiar with *Orobanche* and even commented on the plant's parasitic nature. In northern Europe, early herbalists rarely differentiated species of *Orobanche*, and the woodcut illustrations they provided often lack features that are critical for identification. *Orobanche rapum-genistae*, the species from which the genus derives its common name, was a familiar plant in Britain: as early as 1548 William Turner recorded it in Northumberland as "newe chappel floure" (Britten, 1881). It was well known to Gerard (1597) who lists it in his famous herbal with its host, as though it was a curious manifestation of that plant.

Later taxonomists then grappled with the relationships among these morphologically highly reduced plants. Linnaeus, the 'Father of Taxonomy', who devised the modern binomial taxonomic system in the mid-eighteenth century, only recognised five species in *Species Plantarum* (Linnaeus, 1753). Based in Sweden, he was unfamiliar with the living plants, but his names such as '*O. major*' have long been the source of confusion. He clearly lacked knowledge of their distinguishing features; in fact none of the Father of Taxonomy's proposed *Orobanche* species are now accepted!

The first global monograph of the family Orobanchaceae (as then circumscribed) was written by the Austrian botanist Günther Beck von Mannagetta (Beck, 1890). His student, Alexander von Gilli, continued to work on the genus *Orobanche* in Europe (Gilli, 1966, 1974), Turkey and Asia into the 1970s (Gilli, 1982), but was not ultimately responsible for the detailed account in *Flora Europea*, where a very conservative broader specific concept was ultimately adopted by David Webb and Arthur Chater (Chater & Webb, 1972).

Orobanche variegata
Sicily, Italy (page 16)
photo : Uwe Grabner
Orobanche picridis
South Wales, UK (left)
yellow form (left) and typical form (right)

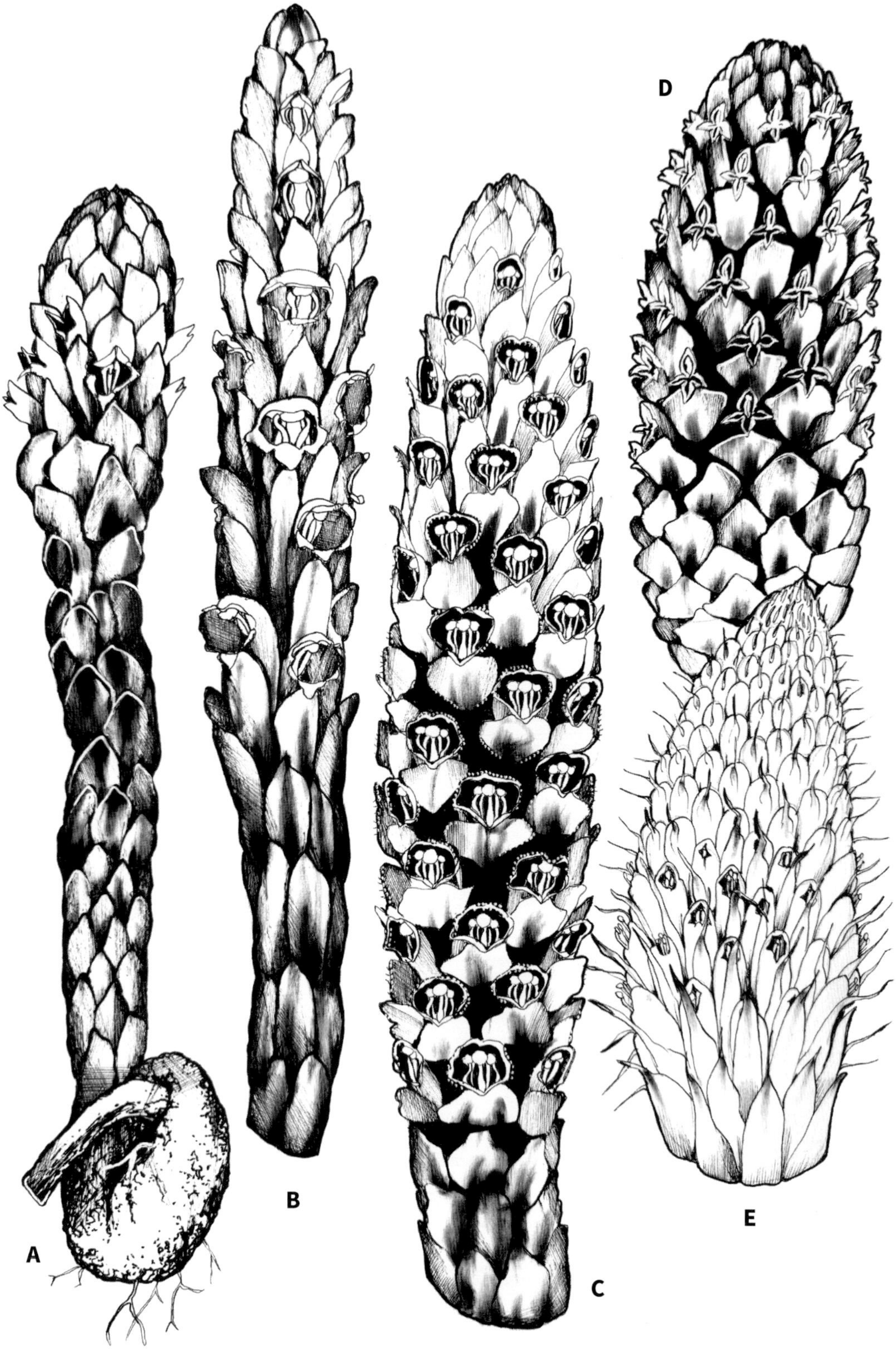

A. *Kopsiopsis hookeri*; **B.** *Xylanche himalaica*; **C.** *Boschniakia rossica*; **D.** *Kopsiopsis strobilacea*; **E.** *Conopholis alpina.*

A–K. *Orobanche* species (inflorescences): **A.** *O. teucrii*; **B.** *O. lucorum*; **C.** *O. reticulata* (*pallidiflora* form); **D.** *O. reticulata* (typical form); **E.** *O. rapum-genistae*; **F.** *O. elatior*; **G.** *O. cernua*; **H.** *O. alba*; **I.** *O. centaurina*; **J.** *O. lutea*; **K.** *O. caryophyllacea*. **L–P.** Corollas in profile (above) and cross section (below); **L.** *O. elatior*; **M.** *O. lucorum*; **N.** *O. caryophyllacea*; **O.** *O. rapum-genistae*; **P.** *O. alba*.

The Orobanchaceae, in its current circumscription, is the largest parasitic plant family, with 102 genera and over 2,100 species of annual and perennial herbs and shrubs (Nickrent, 2020). Except for three genera (*Lindenbergia, Rehmannia* and *Triaenophora*), which are now included in the family, all are parasitic. As discussed above, the family includes both hemi- and holoparasites, and only the latter are covered in this book.

The current circumscription of the Orobanchaceae follows the disintegration and reassessment of the figwort family (Scrophulariaceae) at the end of the twentieth century, as determined with molecular data (Olmstead et al., 2001). This work transformed our understanding of relationships in the Orobanchaceae (Schneeweiss et al., 2004a; Li et al., 2019). The family formerly comprised exclusively holoparasitic *Orobanche* and a few related genera (Beck, 1930); however, on the basis of DNA evidence, it was expanded to include all hemiparasites and the few remaining holoparasites (such as *Lathraea*) that were placed formerly in the Scrophulariaceae (Young et al., 1999; Wolfe et al., 2005; Bennett & Mathews, 2006; McNeal et al., 2013). The sister group to the parasitic Orobanchaceae has been shown by various studies to be the Asian non-parasitic genus *Lindenbergia*, now also included frequently in the broadened Orobanchaceae (Young et al., 1999; Wolfe et al., 2005; Bennett & Mathews, 2006; Park et al., 2008; McNeal et al., 2013).

Since this work, holoparasitic Orobanchaceae now comprise 19 genera and include ca. 320 accepted species (Sánchez Pedraja et al., 2016; POWO, 2025), although the taxonomic position of some is still unclear. Holoparasitic genera were arranged in clades (sensu McNeal et al., 2013) inferred from previous molecular analyses of Orobanchaceae, namely: clade III *Orobancheae*, the oldest and most numerous holoparasitic clade, which includes: *Aphyllon*, *Boschniakia*, *Cistanche*, *Conopholis*, *Epifagus*, *Eremitilla*, *Gleadovia*, *Kopsiopsis*, *Mannagettaea*, *Orobanche*, *Phelipanche*, *Phelypaea*, *Phacellanthus*, *Xylanche*; clade VI *Buchnereae*, which includes *Aeginetia*, *Christisonia*, *Harveya*, *Hyobanche*; and the youngest clade V *Rhinantheae*, which includes *Lathraea*.

Within the revised Orobanchaceae, how the various genera are treated still differs among authors. Although often still grouped as one genus, *Orobanche*, DNA sequence analysis and cytological differences show that *Phelipanche* forms an independent lineage, and so it is treated as distinct in this guide and not as a sub-genus, or section (*Trionychon*) of *Orobanche*. The two other past sub-genera (*Myzorrhiza* and *Gymnocaulis*) are exclusively American in distribution. They share morphological and cytological features and some host preferences with the genus *Phelipanche*, to which molecular data (see for example McNeal et al., 2013) also link them. Meanwhile the North American genus *Aphyllon* is also lumped with *Orobanche* by some authors. Broomrapes of the New World are divided into two well-supported clades sister to *Phelipanche*; these clades are included by Schneider (2016) in the genus *Aphyllon*.

Beck (1930) dedicated significant focus to the genus *Orobanche*. His intra-sectional classification structures, and those proposed by the various authors, were not all confirmed by internal transcribed spacer (ITS) molecular data in the first major phylogeny produced for the genus (Schneeweiss et al., 2004a). For example, his grex *Speciosae*, which included *O. crenata*, was found to nest within the *O. minor* group (Subsection *Minores*). The DNA data also revealed negligible divergence among some taxa that seem distinct in the field. This suggests that ITS sequence data are unable to resolve all taxa adequately.

A. *Orobanche variegata*; **B.** *O. australis*; **C.** *O. sanguinea*; **D.** *O. foetida*; **E.** *O. densiflora.*

A–M. *Orobanche* species (inflorescences): **A.** *O. chironii*; **B.** *O. amethystea*; **C.** *O. picridis*; **D.** *O. minor* subsp. *minor*; **E.** *O. minor* subsp. *maritima*; **F.** *O. canescens*; **G.** *O. artemisiae-campestris*; **H.** *O. hederae*; **I.** *O. iammonensis*; **J.** *O. litorea*; **K.** *O. calendulae*; **L.** *O. crenata*; **M.** *O. almeriensis*; **N–S.** Corollas in profile: **N.** *O. amethystea*; **O.** *O. hederae*; **P.** *O. picridis*; **Q.** *O. crenata*; **R.** *O. minor* subsp. *maritima*; **S.** *O. artemisiae-campestris.*

A–J. *Phelipanche* species (inflorescences): **A.** *P. lavandulacea*; **B.** *P. schultzii*; **C.** *P. resedarum*; **D.** *P. mutelii*; **E.** *P. camphorosmae*; **F.** *P. olbiensis*; **G.** *P. purpurea*; **H.** *P. ramosa*; **I.** *P. arenaria*; **J.** *P. reuteriana*. **K–L.** Corollas in profile (above) and cross section (below): **K.** *P. ramosa*; **L.** *P. purpurea.*

Other sequence data have revealed genetic divergence among closely related taxa in the complex Subsection *Minores* (Thorogood et al., 2008; 2009a, b). Here, the significance of co-evolution and host-mediated speciation, corroborated by detailed molecular studies alongside cultivation experiments, has revealed considerable cryptic diversity that traditional field-based taxonomic methods struggle to accommodate (Thorogood et al., 2008). On the other hand, host identity has been shown to have a strong influence on the morphology of *O. crenata* under controlled conditions (Domina, 2018). British populations of *O. minor* subsp. *maritima* have been found to be closely related to the Mediterranean coastal species *O. litorea.* This hints at the existence of a widespread coastal lineage made up of host-specific cryptic entities; but in the absence of a broadly sampled, well-resolved phylogeny, such relationships remain undetermined. Advanced sequencing DNA technologies may resolve such challenges as those described above.

The extensive diversity of the tribe *Orobancheae*, the oldest and most species-rich lineage of holoparasitic Orobanchaceae, is concentrated mainly in the western Asia and Mediterranean regions of the Old World (McNeal et al., 2013; Schneider & Moore, 2017; Piwowarczyk et al., 2021). Recent investigations have proved that western Asia (particularly the Caucasus) and the Mediterranean are the centre of origin for large clades of holoparasitic *Orobancheae* within the past 6 million years (Piwowarczyk et al., 2021). Many of these species were described only recently, and have not been featured in books before this one.

In the Mediterranean, Asia Minor and the Middle East – the centre of diversity for the genus *Orobanche* – authors have tended to treat newly identified host-specific races with restricted geographic ranges as distinct species. Few of these have been included in phylogenetic analyses, and their genetic relatedness remains unexamined. A combined approach encompassing broad-scale sampling, careful examination of micro-morphological features, and ecological (host) data, as well as advanced DNA sequencing technologies will be required to tease apart relatedness in the most challenging taxa. Until we reach such a stage, taxonomy in the genus *Orobanche* is likely to remain in a state of some flux.

The small genus *Phelypaea*, containing striking plants with single large scarlet flowers, shows greater affinities to *Orobanche* s.str. than that does to *Phelipanche* (Schneeweiss et al., 2004b). Further divisions within *Orobanche* have also been proposed. For example, Carlon et al. (2008) have suggested that on morphological and phylogenetic grounds, the rather distinctive polyploid Mediterranean parasite of *Salvia rosmarinus*, *Orobanche latisquama* should be treated in its own genus *Boulardia*. Meanwhile, the monotypic genus *Necranthus*, created by Gilli in 1968 based on the Turkish *N. orobanchoides*, has been shown to belong to *Orobanche* s.str.

Cistanche is a diverse genus of robust parasites (sometimes called 'desert broomrapes' or 'desert hyacinths'), which typically parasitise the roots of small trees and shrubs. These conspicuous plants are widespread across dunes, deserts and semi-arid regions of the Old World. Our understanding of species diversity in the genus *Cistanche* is still rudimentary and in a state of flux.

Toothworts (*Lathraea*) is a further genus of holoparasites in the family, distantly related to *Orobanche*, and resolved within a lineage shared with *Euphrasia*, *Parentucellia*, *Odontites* and *Rhinanthus* (Li et al., 2019). This small genus is typically parasitic on the roots of trees in forested habitats across temperate Europe and Asia.

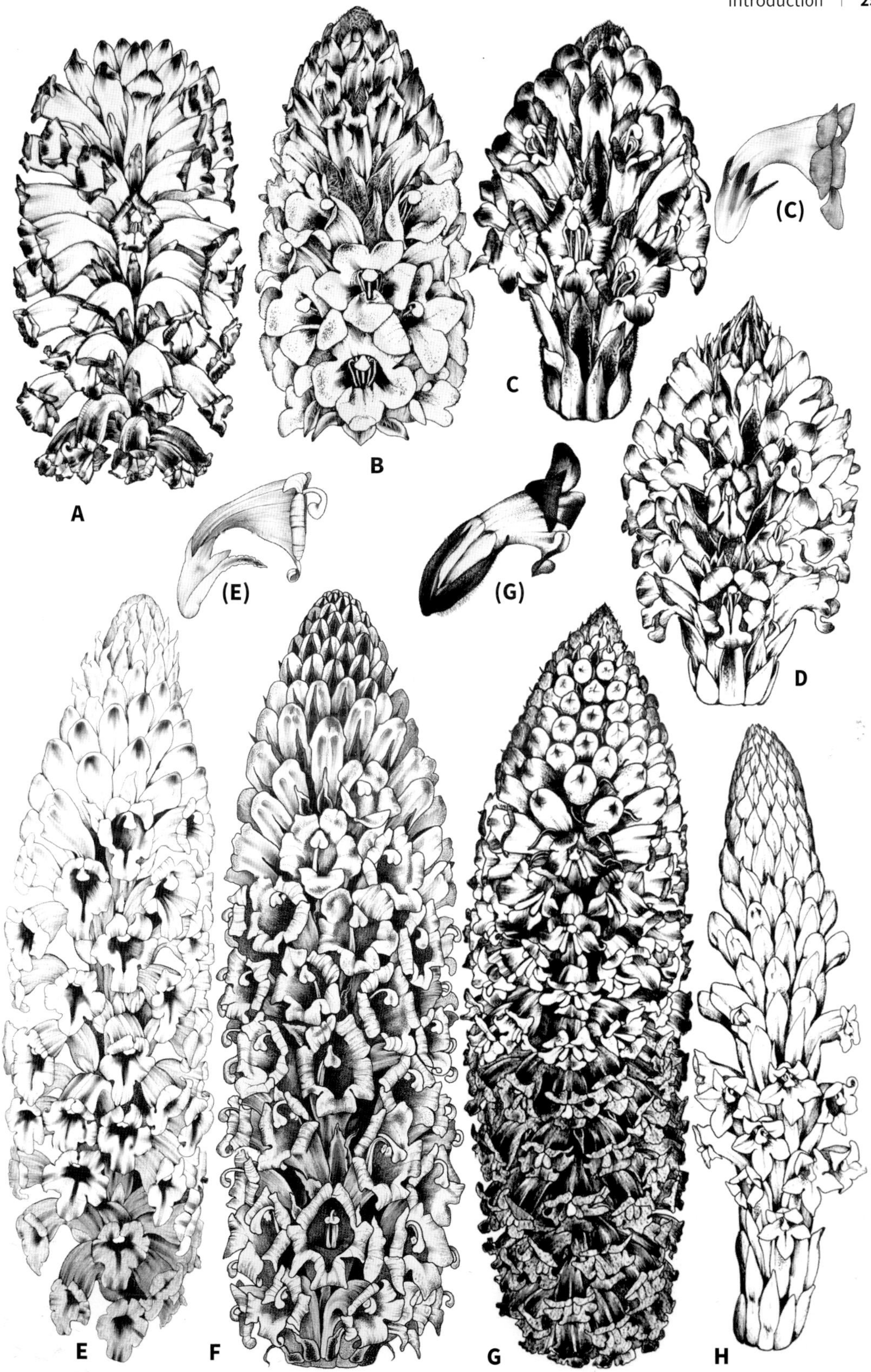

Cistanche species (inflorescences – letters without parentheses; and corollas in profile – letters in parentheses): **A.** *C. violacea*; **B.** *C. mimii*; **C.** *C. salsa*; **D.** *C. sinensis*; **E.** *C. phelypaea*; **F.** *C. tubulosa*; **G.** *C. deserticola*; **H.** *C. laxiflora*.

The genus *Lathraea*: **A.** *L. rhodopea*; **B.** *L. squamaria*; **C.** *L. purpurea*, **D.** *L. japonica*, **E.** *L. clandestina*.
Adapted from Hatt et al., 2024, reproduced with permission from the copyright holder (Magnolia Press).

In this book we focus on flower morphology, host specificity and ecology as key diagnostics for the various species and genera. In practice, it is often useful to consider a range of variables, especially when identifying closely related species. Seed ornamentation and morphology are not generally considered to be useful traits for distinguishing broomrape genera (and in any case, require scanning electron microscopy, so are useless in the field), although data describing seeds from across the family are scarce (Piwowarczyk, 2015a, b; Piwowarczyk et al., 2020a).

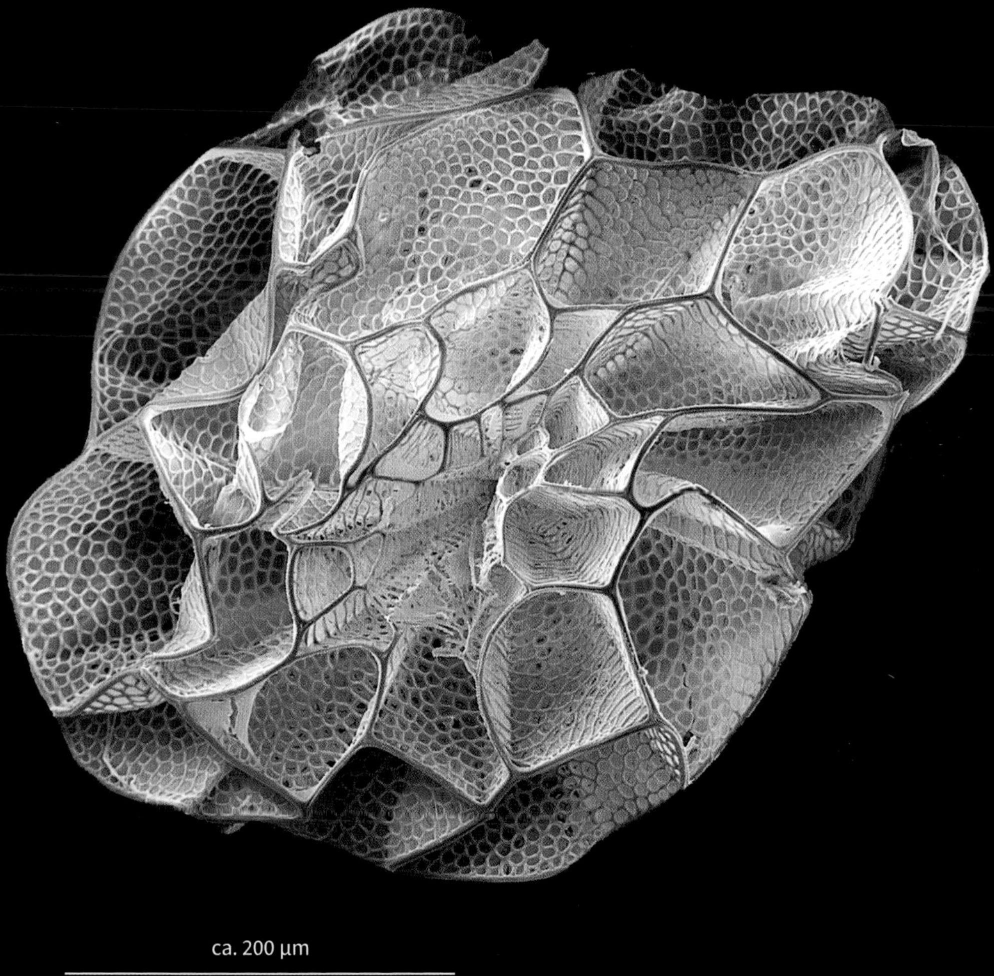

Xylanche himalaica seed seen under scanning electron microscopy.
Seed from Nepal, photo: Justyna Kasińska.

A. *Orobanche reticulata* (photo: Grahame Preston); **B.** *O. colorata*; **C.** *O. picridis*; **D.** *Cistanche mauritanica* (photo: Mimoun Bachiri); **E.** *O. minor* f. *lutea*; **F.** *C. deserticola*; **G.** *Harveya capensis* (photo: Mayur Prag); **H.** *O. cypria* (photo: Charalambos S. Christodoulou).

F
G
H

A
B
C
D
E
F

1

LIFE CYCLE AND BIOLOGY

Broomrapes are all root parasites but their life cycles and biology differ with their various hosts, distributions and ecologies. The visible portion of a broomrape that emerges above ground typically comprises a flowering spike or a sessile cluster of flowers (or in some cases a solitary flower). The plant is devoid of functional leaves (these are typically reduced to scales at the base of the stem). The life history of most species – indeed many genera – is poorly known or undocumented, particularly with respect to their longevity. The most well-studied genus is *Orobanche*, for which a brief account is given here.

(page 30) The life cycle of a broomrape (*Orobanche minor*). **A.** Seeds; **B.** Seeds germinate in response to host-derived chemical signals; **C.** Seedlings attach to host root; **D.** Tubercle formation; **E.** Host-derived starch accumulation ('spider' stage – in annual species); **F.** Inflorescence development, followed by seed production.

(left) An excavated specimen of *Orobanche coerulescens* attached to the roots of its *Artemisia* host. **H** = Haustoria (the point of connection); **HR** = Host Root.

The life cycle of a broomrape above ground (*Cistanche phelypaea*): **A–B.** Emergent buds; **C–D.** Emergent inflorescences (flower formation); **E.** Seed production and dehiscence

B
C
D
E

Many *Orobanche* species are annuals and are able to germinate and set seed in the same year, for example the weedy species *O. crenata*, which often occurs as a weed in bean fields and allotments in the Mediterranean and North Africa. Some species are perennial – often weakly so; indeed, fresh flowering spikes can be seen occasionally growing alongside withered spikes from the previous year, for example *O. rapum-genistae*. Rather more commonly, broomrapes are monocarpic, dying after flowering. For example, *O. hederae* flowers two years after germinating but seldom shows unequivocal evidence of perennation. The life history and longevity of broomrapes is also influenced by, or coordinated with, their host plants' physiology. *Cistanche* seems to have a similar life cycle, albeit in desert or other arid habitats. *Lathraea*, however, propagates via rhizomes around the roots of trees and is demonstrably perennial and long-lived (populations persist for decades).

Broomrape seeds can remain viable after considerable periods of dormancy, in some instances for 20+ years. Those examined (especially *Orobanche*) germinate only following a conditioning period of suitable temperature, and in response to chemical signals produced by host plants' roots. These signals include strigolactones, which are exuded by the roots to signal to symbiotic arbuscular mycorrhizas in the soil. Differences in signal responsiveness and perception by the parasite, as well as differences in signal production by the host, influence the level of parasite germination and establishment. Host signals also govern the later stages of the host–parasite interaction; in a successful interaction, the radicle of the broomrape seedling attaches to the host root via a haustorium – this is the physiological bridge via which the parasite intercepts water and nutrients. On a compatible host, the parasite develops either tubercles, fleshy root-like starch-storage structures, or rhizomes, depending on the genus. At maturity, an inflorescence is produced – typically in spring and summer in temperate biomes. The dried fruiting spikes are often woody and can persist long after the capsules have released their seeds. The phenology of tropical and subtropical genera such as *Christisonia*, which occurs from the Indian subcontinent to the Philippines, is largely unknown.

Cistanche tubulosa in the deserts of the West Bank, Palestine (page 34)
Orobanche picridis, visited by a bumblebee, UK (left)

A
B
C
D
E
F
G

2

MORPHOLOGY

Broomrapes possess peculiar morphological structures. Many have fleshy stems with leaves reduced to scales, roots replaced by haustoria, and sometimes extensive underground rhizome systems (especially in *Cistanche* and *Hyobanche*). Their morphology ranges from geoflorous plants a few centimetres high (e.g. in *Hyobanche*) to robust clumps of plants over a metre (*Cistanche*). Most have a characteristic fleshy appearance, bright colours, and, to varying degrees, are covered in trichomes (hairs).

Broomrape identification and taxonomy are notoriously challenging. While it is not always possible to identify closely related species in herbaria or from photographs, with careful observation, identification in the field can be straightforward for most broomrapes.

Populations of broomrapes often vary in robustness and colouration so care should be taken not to rely on a single diagnostic character. In *Orobanche* (and probably other genera), characters that are especially important to observe include the colour and shape of the corolla; the presence or absence of dark glands on the corolla; the colour and degree of fusion of the stigma lobes; the position, shape and size of the lobes of the upper and lower corolla lips, and the presence or absence of glands on their margins; the length, shape and venation of the calyx (though this may vary even on a single plant), and the extent to which the calyx segments are fused together (connate) under the flower; the height of insertion of the stamens (i.e. the distance from the base of the corolla tube at which the filament becomes free); the position and degree of the stamen hairiness; and the level of glandular hairiness. The nature and extent of hairiness and the extent and nature of the swelling of the shoot base are also useful characters.

The presence of bracteoles is an easy way to distinguish *Phelipanche* from *Orobanche* (species in the former genus are also typically bluish or purplish-mauve in colour). In fruit, the two genera can easily be distinguished as the capsules of *Phelipanche* do not remain fused at the apex as they do in *Orobanche*. Similar diagnostics are important for the genus *Cistanche,* but the presence or absence of pubescence is crucial for this genus. For most of the smaller genera, *Lathraea* for example, there are so few species that identification is almost always straightforward.

(left) **A.** *Cistanche tubulosa*; **B.** *C. violacea*;
C. *Harveya hyobanchoides* (photo: Abdul-lateef Ismail);
D. *Orobanche flava*; **E.** *O. elatior*; **F.** *O. densiflora*
(photo: Gonzalo Moreno Moral); **G.** *O. foetida.*

B
C
D
E
A
F
G

Many broomrapes have a restricted host range, as discussed above. Identifying the host, as well as examining morphological features, is thus helpful where possible. For some species this is straightforward: *O. rapum-genistae* will always be found close to its host – a woody Fabaceae species such as *Ulex* or *Cytisus*. For species with a broad host range, such as *O. minor*, reliable identification of the host can be all but impossible when the plants are found in mixed vegetation. An overlap in the host range of different species can also present difficulties. *Orobanche minor* can grow on ivy (though rarely), which is of course the predominant host of the similar species *O. hederae*. For these reasons, host plant identification alone is never a reliable diagnostic.

Perhaps the most morphologically challenging group among all broomrapes is *Orobanche* Subsection (grex) *Minores* (Beck) Teryokhin (Teryokhin et al., 1993). This is a group of similar, small-flowered species (with corollas typically <20 mm) that are notoriously challenging to identify in the field, and that preserve poorly in herbaria. Here, host-defined forms of broomrape seem to be in the process of incipient speciation – i.e. they are on the evolutionary path to forming new species.

Through researching this book, we estimate that there are ca. 320 holoparasitic species of Orobanchaceae worldwide. In this guide we include descriptions of 221 species (and a further 25 infraspecific taxa) representing each of the main 19 genera, but there is an inevitable bias towards the more speciose, especially the genus *Orobanche*, which alone contains ca. 120 known and accepted species (see, for example, Index of Orobanchaceae), or in its broadest treatment, i.e. including *Phelipanche* and other closely related genera, ca. 200 species (POWO). Some of these species are very poorly known. We use the most widely accepted scientific names, but we also include some of the more common synonyms to avoid confusion; it would be impossible to list all synonyms for practical reasons.

(left) The anatomy of a broomrape: **A.** Inflorescence; **B.** Corolla in profile; **C.** Corolla in cross section; **D.** Stigma (and style base); **E.** Carpel (with basal ovary, style and stigma); **F.** Floral bract; **G.** Calyx segment.

3

REPRODUCTION

Broomrapes have evolved a suite of adaptations to attract pollinators. Chasmogamous flowers (those that are open to promote cross-pollination) are common; cleistogamous flowers (those that are closed and self-pollinating) also occur in the genera *Epifagus* and *Boschniakia* (Thieret, 1971; Musselman, 1980). Some are facultatively selfing, such as species in the genus *Orobanche* Subsection *Minores.* Typically, broomrape flowers are campanulate or tubular, zygomorphic (bilaterally symmetrical) with a 2-lipped corolla. They possess a diversity of pollinator attraction mechanisms from conspicuous visual and tactile nectar guides and rewards to UV-absorption patterns, and morphological features such as anther-like swollen folds or dumbbell-shaped stigmas, as well as scents (from clove- to carrion-scented) arising from volatile organic compound (VOC) signatures, and they have unique stigma microbiota (Tóth et al., 2016; Piwowarczyk & Kasińska, 2017; Ruraż et al., 2023; Lunau et al., 2024).

Broomrapes produce a diversity of forms from small flowers packed in dense spikes, for example in *Orobanche*, *Phelipanche*, *Hyobanche*, *Cistanche*, to single-flowered inflorescences (e.g. *Phelypaea*, *Aeginetia*, some *Aphyllon*), many of which are conspicuous with an open throat, bright colouration, and attractive to insects. Pollinators include a range of arthropods, especially Hymenoptera, including bumblebees, bees and wasps; records are scant for most species. Some species, from subtropical and tropical regions, such as some *Hyobanche*, *Harveya* and *Orobanche boninsimae*, are pollinated by mammals or birds (Turner & Midgley, 2016; Niedzwetzki-Taubert, 2022; Nishimura & Takayama, 2022). *Harveya speciosa* is pollinated by long-tongued hawkmoths (Johnson & Raguso, 2016); we have also observed hawkmoths visiting *Cistanche*, but data on such observations are scarce. Some, such as *Orobanche minor* and closely related species, appear to be predominantly self-pollinated, at least in the north of their range (northern Europe).

The fruiting spikes of broomrapes can produce many thousands of minute seeds (some of similar size to those of orchids). In *Orobanche*, these are often wind- or water-dispersed following the release from their capsules. The large quantity of seeds, their (typically) wind dispersal, and their ability to be washed into crevices, together maximise opportunities for the parasites to encounter the roots of a suitable host plant. Mainly self-pollinating species such as *O. minor* will always set seed. Others, particularly those with larger, fragrant flowers, are predominantly outcrossing and appear to set rather smaller quantities of seed (or in some specimens, such as *O. caryophyllacea*, no seed at all).

Cistanche phelypaea visited by *Hyles livornica*
Fuerteventura, Canary Islands (left)

Hybrids

There is little documented evidence of hybridisation in the Orobanchaceae. Many genera, including the largest, *Orobanche*, contain predominantly monocarpic annuals or short-lived perennials and have no means of vegetative propagation. Any hybrids that might arise would therefore have limited opportunity for dispersal were they to be fertile, and as and when they did arise, they would be ephemeral only. Even in perennials such as *Lathraea*, hybrids are unreported. Nevertheless, hybrids could have been under-recorded in the Orobanchaceae because they are likely to be overlooked given the taxonomic complexity of, and cryptic speciation within, the genus. For example, all *Cistanche* species appear to be insect-pollinated and outcrossing, but species limits are so poorly known that hybrids are likely to go unnoticed.

This may be the case even in better-understood groups. For example, returning to *Orobanche*, species in the *Minores* group (*O. minor* and its various infraspecific taxa, *O. picridis* and *O. hederae*) are so similar that any hybrids would probably go unnoticed (Thorogood & Rumsey, 2020). Morphological intermediates among these taxa, for example subsp. *maritima* and subsp. *minor*, do occur in southern Britain (Cornwall, the Scillies and the Channel Islands), where the former shows a broader than normal host and habitat range, as well as at the eastern end of its range on the chalk cliffs and dunes of Kent (south-east England). The authors have also found hybrids of *O. minor* and *O. picridis* in Wales where these species co-occur; some of these hybrids yield no seed and appear to be sterile. Artificial hybrids have also been raised in *Orobanche*, for example between *O. minor* and *O. hederae* (Thorogood & Rumsey, 2020), and these species also co-occur occasionally where their habitats merge in Europe, so the possibility of natural hybridisation in these species is not out of the question either.

More broadly among broomrapes, hybridisation might be expected to be rare because species are often allopatric (by virtue of their hosts' ecologies). Any hybrids that do arise may be less well-adapted to a given host plant than their parent species and outcompeted by them. Polyploidy is unevenly distributed in *Orobanche* – the largest of the genera – and restricted to a few lineages, for example *O. gracilis* and its relatives; in related genera, polyploids are rare or absent (Schneeweiss et al., 2004b). Allopolyploidy is the prevalent mechanism for speciation in many plant groups. The low incidence of hybridisation here, for the reasons given, means that this is not the case for broomrapes.

The natural hybrid of *Orobanche picridis* × *O. minor* South Wales, UK (page 42)

Orobanche minor (left) and *O. picridis* (right), showing the close morphological similarities of these closely related but distinct species. South Wales, UK (left)

4

HABITATS, ECOLOGY AND DISTRIBUTION

Broomrapes' habitats and ecologies are much defined by their host range. Habitats range from extreme deserts (for example the genus *Cistanche*) to closed canopy temperate forests (for example *Phacellanthus* in Japan) and mountains (for example *Gleadovia* in the Himalaya). Each species of broomrape has a preferred host or group of hosts, for which the underlying determinants are complex and multifaceted. The success or failure of the various developmental stages outlined above influences the host specificity of the parasite. The development of the parasite is also determined by other host attributes, such as root architecture and spread, and growth rate.

A broomrape will often appear some distance from its host plant; the host can only be identified with certainty by excavation and can be difficult to determine in mixed vegetation. While some broomrapes are highly generalist and can infect numerous hosts from unrelated families, from annuals or perennials to vines, shrubs and trees, others are restricted to a single species. Moreover, even host generalists can comprise of multiple more host-specific taxa. *Orobanche minor* has a very broad host range and can parasitise hundreds of host species from many families. However, this broomrape shows a preference for particular families, such as Fabaceae, Asteraceae and Apiaceae. Moreover, races of this species (which are also recognised at the infraspecific rank) are also specific to, for example, sea carrot (*Daucus carota* subsp. *gummifer*), *Brachyglottis* (a widely planted ornamental shrub native to New Zealand) and sea holly (*Eryngium maritimum*). By virtue of these hosts' different niches – namely sea cliffs, urban landscapes and coastal dunes – the various forms of *O. minor* vary in their ecology accordingly.

The rather exacting requirements for a particular habitat and host plant, and the requirement for at least some soil disturbance to facilitate host root–seed contact, probably explains why many broomrapes are rare, infrequent or local. Some have retracted markedly within their former range, for example *O. rapum-genistae* in northern Europe, now much reduced; others appear to have expanded, for example *O. crenata,* a Mediterranean species that has naturalised with some permanence as far north as southern Britain, and *O. minor,* which is now a cosmopolitan weed. Others are widespread geographically – perhaps owing to wind-dispersed seed and widespread hosts – but may be local at their range edge. For example *O. caryophyllacea* extends south to Morocco and east to Kazakhstan, Pakistan, Tajikistan and Transcaucasia where it predominantly occurs in mountains; but at its northernmost limit in Holland and south-east Britain it is rare and restricted to low-lying habitats such as coastal dunes. Similarly, *O. picridis* has a predominantly southerly distribution, extending to Asia Minor and North Africa; its restriction to the far south of Britain in dunes and on cliffs indicates a mere foothold into the northernmost extremity of its range.

(left) Semi desert hills in Azerbaijan, the habitat of *Cistanche fissa* (inset)

Centres of diversity vary among the genera. The genera *Orobanche* (ca. 120 species) and *Phelipanche* (ca. 60 species) are most diverse and occur worldwide, mostly in temperate regions of Europe, especially in the Mediterranean Basin, western and central Asia, northern Africa, and although less represented (and probably introduced), in the Americas and Oceania. The genus *Aphyllon* is sister to the Old World genus *Phelipanche*, and contains over 20 species that are restricted to the New World, especially North America. The genus *Cistanche* (ca. 25 species), which occurs mainly in dunes, semi-desert and desert habitats, is distributed from Macaronesia to China, with an apparent centre of diversity in the Middle East; but again, in the absence of a well-sampled, resolved phylogeny, this is difficult to ascertain (a challenge for all genera). Southern Africa is a centre of diversity for the genera *Harveya* with ca. 28 species, also distributed from Eritrea and Madagascar and the Mascarene Islands, to the Arabian Peninsula, north to central Asia, and *Hyobanche*, with eight species that occur mainly on sandstone fynbos and are endemic to southern Africa from the Cape Floristic Region northward to Namibia. *Christisonia* comprises ca. 24 species in tropical forests in tropical Asia from India, south to South-east Asia, and east to southern China and the Philippines. *Aeginetia* includes seven species found mainly in rainforests of south-east and East Asia.

A few genera are very small, at least in their current circumscription, for example *Phelypaea*, with three species native to the Caucasus, the Middle East, and more rarely in Crimea and the Balkans; *Lathraea* includes five accepted species, with a general native range in temperate Eurasia. *Boschniakia* s.l. includes four species that differ in their geographical distribution and are currently divided in various genera: *Boschniakia* (one species), *Kopsiopsis* (two species) and *Xylanche* (one species). The range of these species extends into the farthest north and in cold areas at high altitudes, such as in woods and tundra, on rocky slopes or cliffs, mainly in western North America and extreme north-eastern Asia and Himalaya. *Gleadovia* (five species) occur in forests from Himalaya to south China. Monotypic *Phacellanthus* is distributed in the woods of eastern China, Taiwan, Japan, Korea and Russia. *Mannagettaea*, with two mostly forest-dwelling species, is known from north-central and south-central China, Irkutsk, Qinghai and Tuva. Small New World genera include *Epifagus* (one species) and *Conopholis* (two species), which occur mainly in the temperate forests in north-eastern North America, and the monotypic *Eremitilla*, described recently from montane forests in south-west Mexico.

Some species in the Orobanchaceae have a profound effect on both natural and agricultural systems. For example the hemiparasitic yellow rattle (*Rhinanthus minor*) parasitises grasses, supressing their dominance, so enabling a diversity of other flora to prosper. Meanwhile striga (*Striga*) destroys billions of dollars' worth of cereals and legumes in Africa and Asia every year. Some species of broomrape (*Orobanche* and *Phelipanche*) are also pernicious weeds of various crop species in the Mediterranean Basin and the Middle East.

(right) Mediterranean coastal dunes on Elafonisi Island, Greece, the habitat of *Orobanche sanguinea* (inset)

(page 48–49) Temperate mountain forest in Hokkaido, Japan, the habitat of *Phacellanthus tubiflorus* (inset)

(page 50–51) Temperate mountains in Turkey, the habitat of *Phelypaea boissieri* f. *lutea* (inset) photos: Ümit Subaşı

5

TRADITIONAL USES

Broomrapes have a rich history of traditional use. Although the phytochemistry of most species is unknown, the plants are established to be rich in phenylethanoid glycosides – plant-based compounds that have been used in traditional medicine. Broomrapes derive all nutrients from their hosts, but they maintain their own biosynthetic pathway and thus phytochemical profiles (Piwowarczyk et al., 2020b; Lachowicz-Wiśniewska et al., 2024). While some parasitic plants are pests on crops, others are cultivated in their own right for food, industrial or medicinal products (Pignone & Hammer, 2016; Těšitel et al., 2021). Broomrapes have been used for thousands of years in traditional medicine, especially in China and America, as herbal or functional supplements (Shi et al., 2020).

The fleshy underground stems of *Cistanche*, known locally as 'Rou Cong Rong' or 'desert ginseng', have been used in traditional Chinese medicine for more than 2,000 years (Li et al., 2016; Thorogood et al., 2021). The plant is used as a tonic herb, and extensive pharmacological and chemical reports have been published, which identified some 100 phytochemical compounds including phenylethanoid glycosides, lignans, alditols and iridoids that together have a range of pharmacological properties (Fu et al., 2018). Unfortunately, many *Cistanche* are rare or endangered and threatened from excessive harvesting or illegal trade. In some areas of Puglia in Italy, *O. crenata* is consumed in a similar way to asparagus, and widely used in traditional dishes; as a vegetable, the plant has interesting nutritional benefits including a high fibre content, antioxidant capacity and a high level of polyphenols (Renna et al., 2015). In South Africa, *Harveya* and *Hyobanche* species are commonly known as '*inkblom*', because they turn black upon drying, and early settlers used the flowers to make ink. In Armenia, *Phelypaea tournefortii* flowers have been used to dye foods (e.g. eggs) and wool, due to their significant quantity of red anthocyanin pigments (Piwowarczyk et al., 2020c). *Aeginetia indica* is used as a food colourant in traditional Thai dishes, and was used historically as a folk medicine (On-Nom et al., 2024).

Underground stem of *Cistanche deserticola*, known as 'Rou Cong Rong' harvested for herbal medicine
Inner Mongolia, China (left)

6

CONSERVATION

Besides a handful of species, few parasitic plants are widespread in cultivation; indeed it has been estimated at least 76% of species are entirely missing from collections today, most of which have never been grown (Thorogood et al., 2021). Parasitic plants are also conspicuously under-represented in seed bank collections and national *ex situ* conservation programmes regionally. Parasitic plants' life cycles are coordinated with those of their hosts, and their successful cultivation relies on an understanding of their development; for many, the biology and ecology are poorly known, and they are probably neglected because of a presumed difficulty in propagation. Moreover, most holoparasitic species are rare and endangered, and completely dependent on their hosts. This dependency is likely to be impacted by climate change, which may affect holoparasites both directly through effects on their physiology, and indirectly via their host plants and habitats (Piwowarczyk & Kolanowska, 2023a, b).

Historically, broomrapes have been planted more as botanical curiosities than for conservation *per se*. The University of Oxford Botanic Garden is the oldest botanic garden in the UK, founded in 1621. It was established as a physic garden in which medicinal plants were grown for teaching purposes. This botanic garden has a long history of parasitic plant cultivation, beginning with *Lathraea squamaria*, which was introduced to the garden during the 1600s; *Orobanche* species new to the UK were later introduced in the 1980s and 1990s (*O. flava* still persists in 2024). Today the garden has a diverse living collection of holoparasitic Orobanchaceae. A substantial collection of parasitic plants was also grown at Marburg Botanic Garden for several decades, in connection with the research of Prof. H.-C. Weber. Unfortunately, this collection vanished with his retirement and the subsequent reorganisation of the botanic garden. Other botanic gardens that are notable for their parasitic plant collections include Copenhagen Botanic Garden, Innsbruck Botanic Garden and Munich Botanic Garden (Heide-Jørgensen, 2008; Thorogood et al., 2021). In 2020, a separate collection of parasitic plants was also established at the Botanic Garden in Kielce (Poland) in cooperation with the Center for Research and Conservation of Biodiversity of the Jan Kochanowski University. This collection currently includes seven nationally endangered species of *Orobanche* and *Phelipanche*. At the same time, in the Center at the Jan Kochanowski University, a seed bank of holoparasitic Orobanchaceae was created to represent mostly species from Europe and western Asia.

Besides *ex situ* propagation and seed-banking efforts, reports of broomrape conservation are all but absent. Botanists at Oxford Botanic Garden have attempted the reintroduction of *Orobanche rapum-genistae* to local reserves in Oxfordshire in the UK where the plant is locally extinct. Such efforts can be resource- and time-expensive. Future effort would benefit from an approach combining taxonomy-informed conservation – as we can only protect what we know to exist – coupled with adequate representation in living collections and seed-banking strategies.

Cistanche deserticola, farmed sustainably, being harvested for herbal medicine Inner Mongolia, China (left)

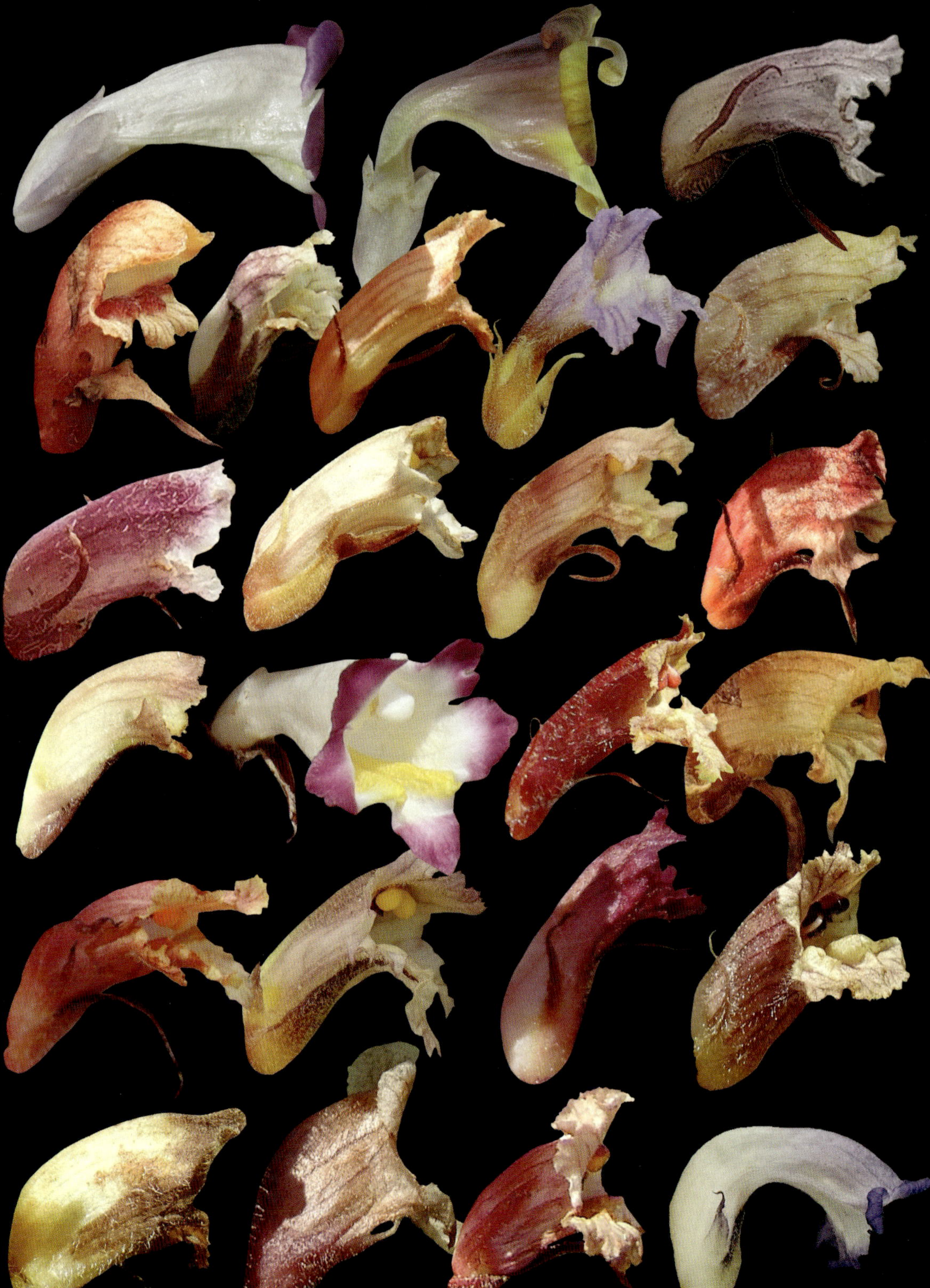

THE SPECIES

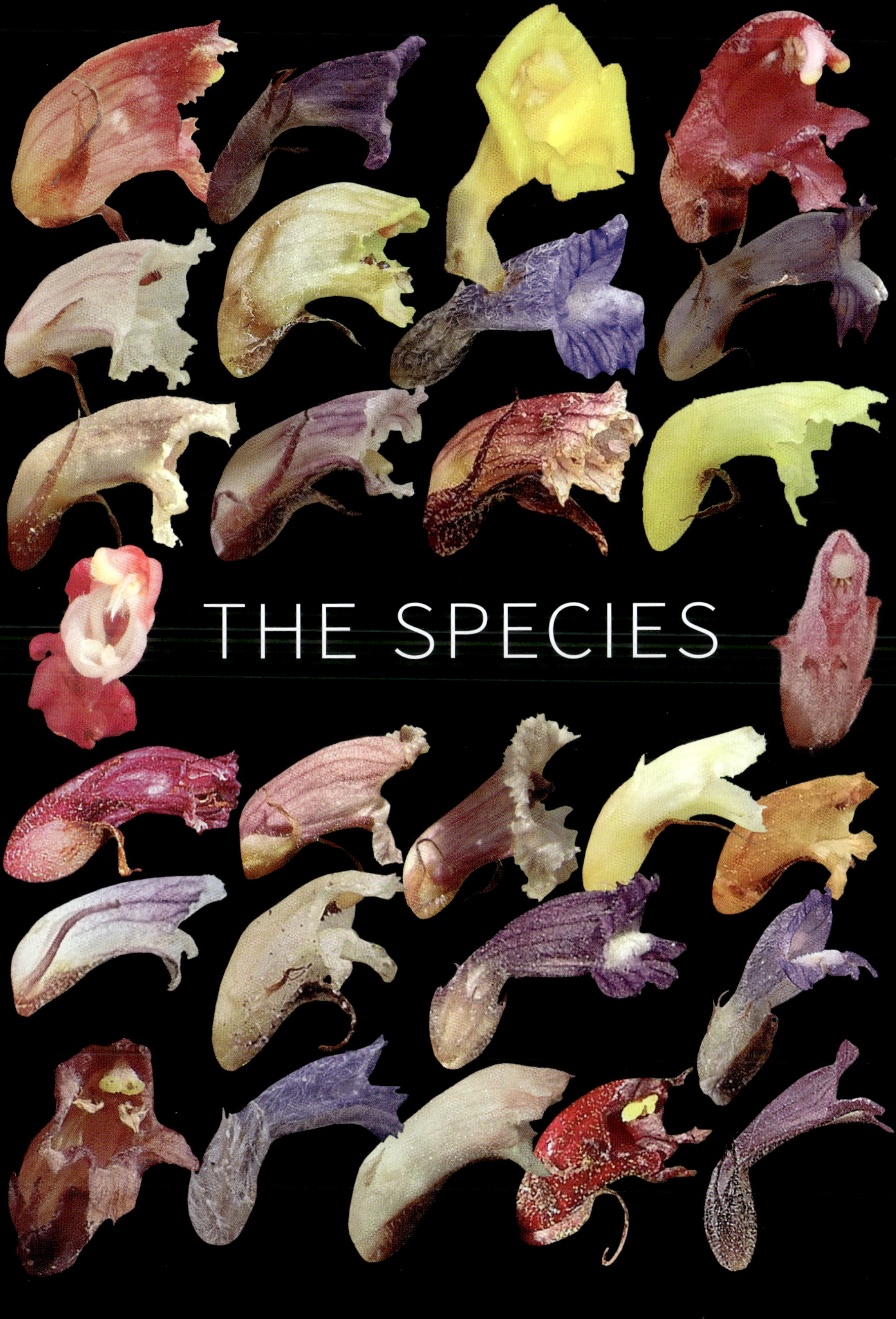

NAME ***Aeginetia acaulis*** (Roxb.) Walp.
SYNONYMS ***Aeginetia abbreviata*** Buch.-Ham. ex Benth.; ***Aeginetia pedunculata*** Wall.; ***Orobanche acaulis*** Roxb.

Habit: clumped, erect. **Height**: 6–14 cm. **Stem**: subterranean stem short, cylindrical, unbranched, glabrous, 2–6 cm. Inflorescences short, racemose, 3–several flowered. Stems cream to reddish, flexuous, branched, rounded, protruding slightly above the ground, 2–3 cm long; scales sparse around the stem base, ovate-triangular, 1 × 0.5 cm, glabrous, the apex acute or slightly obtuse. **Calyx**: spathe-like, unilaterally cleaved, very irregularly bifid; reddish or yellow, subequalling or slightly shorter than the corolla. **Corolla**: ellipsoid in bud; tube smoothly curved (not geniculate), the base constricted. Corolla indistinctly bilabiate with 5 lobes; lobes subequal, circular to reniform, the margins dentate; tube 0.8–1 cm × 5–7 cm long*;* tube pale yellowish-cream externally, the lobes blue to blue-purple; throat yellow, edged deep blue. **Carpel**: stigma inflated, discoid, 0.3–4 cm in diameter, weakly trilobed to shield-shaped; style 2 cm, persistent, curved. **Stamens**: filaments glabrous; anthers white, 0.15–0.2 cm long, uneven. Filaments inserted about ⅔ from the tube base, with a conspicuous tuft of hairs, effectively closing the corolla tube throat. **Host**: Poaceae (various). **Ecology and distribution**: southern China and south-east Asia except Laos; southern Asian tropics. Restricted to deciduous dipterocarp, bamboo forests and grasslands, 50–1,100 m. **Key diagnostics**: clusters of blue-purple flowers with cream calyces. **Similar species**: similar to *A. indica*, but with larger, bluer flowers on shorter pedicels.

Aeginetia acaulis (above, right) photos: Anoop P. Balan

NAME ***Aeginetia flava*** J.Parn.

Habit: clumped, erect. **Height**: 15–20 cm. **Stem**: solitary or branched near the base, dull reddish-yellow, with opposite, narrowly triangular scales, 1 × 0.2 cm. **Calyx**: spathulate, the tip erect to recurved, whitish-yellow at the base grading into pale pinkish at the tip; glossy; 2.5–4 cm long. **Corolla**: ellipsoid in bud; tube exserted 2.5–4 cm beyond the calyx; the exserted portion downward-curved or ± straight, somewhat dilated distally; the concealed part of tube strongly curved to ± geniculate, forming a narrow throat above the ovary, the throat somewhat dilated in its mid-portion. Lobes widely spaced, slightly recurved at tips, markedly but shallowly and irregularly dentate with teeth ca. 0.1 cm; tube 3.5–5.5 cm; lobes ca. 0.5–1 cm; bright butter-yellow, externally glossy. **Carpel**: stigma yellow, included; peltate; style yellow, curved almost to the roof of the corolla tube. **Stamens**: included, massed together, attached ± at point of inflection of the corolla tube; point of attachment hairless; anther dehiscence by ± apical pore-like slits; upper stamens spurred and gibbous, the lower gibbous but only minutely spurred; spur with a terminal conical projection at the tip; filaments of upper stamens distinctly papillate near the anthers; filaments of lower stamens ± smooth, anthers hairless. **Host**: *Strobilanthes* spp. (Acanthaceae). **Ecology and distribution**: south-eastern Thailand in *Quercus semiserrata*-dominated rainforests, 1,400–1,540 m. **Key diagnostics**: clustered, bright yellow flowers. **Similar species**: most similar to the morphologically variable *A. indica*, but with conspicuous basal stem scales, sometimes branched stems, and with bright yellow flowers with dentate lobes.

Aeginetia flava Thailand (above, right) photos: Bob Harwood

NAME ***Aeginetia indica*** L.
SYNONYMS ***Orobanche aeginetia*** L.; ***Phelypaea indica*** (L.) Spreng. ex Steud.

Habit: clumped, often with numerous ascending slender stems. **Height**: 15–40(–50) cm. **Stem**: solitary, slender, unbranched (rarely branched at the base), glabrous; yellow, pale brownish-red, maroon, or red-striped; to 30 cm long, each with a single terminal flower. Scales subterranean, red, ovate-lanceolate, or lanceolate, 0.5–1 × 0.3–0.4 cm. **Calyx**: spathe-like, split in front nearly to the base; tip erect to recurved, purple, pink, rarely white; 2.5–3.5 cm long. **Corolla**: indistinctly bilabiate, tubular-campanulate; tube slightly curved, lobes subentire; limb obscurely 2-lipped with 5 subequal lobes, pinkish-purple, striated; exserted region of the tube usually purple-maroon or occasionally white; concealed tube strongly reflexed, often ± geniculate, forming a narrow throat above the ovary; 2–4.5 cm; tube whitish, variably suffused purple-red, to deep maroon. **Carpel**: stigma pale yellow or whitish-pink, conspicuous, peltate; style slender, 1–1.5 cm, yellow or whitish-pink, curved. **Stamens**: inserted, filaments glabrous, purple; anthers yellow, 0.7–0.9 cm; point of filament insertion hairless. **Host**: various monocots in the families Poaceae, Cannaceae, Cyperaceae, Dioscoreaceae, Juncaceae and Zingiberaceae. **Ecology and distribution**: south and south-east Asia, including Java, Philippines, Borneo, Japan, and Papua New Guinea. Typically in evergreen, bamboo, dipterocarp and pine forests with granite or limestone substrates or grassy lowlands, or alluvial areas, 0–1,600 m. **Key diagnostics**: clumped stems bearing solitary, somewhat nodding flowers with spathe-like calyces. **Similar species**: similar to *A. flava* but lacking conspicuous scales at the stem base above-ground. Flowers of *A. flava* are yellow with broadly spreading dentate lobes, whereas in *A. indica*, the corolla is purplish, the lobes not dentate, and scarcely spreading.

Additional species

NAME ***Aeginetia mirabilis*** (Blume) Bakh.
Recognised by its rather short, stout stems, scarcely raised above-ground, 1.5–5 cm long, 0.3–0.5 cm in diameter, red-rust coloured, with rather dense, boat-like scales; corolla dark violet; calyx with a long apical spire. Apparently parasitic on Poaceae, Moraceae, Zingiberaceae and Piperaceae. Native to Java.

NAME ***Aeginetia mpomii*** Letouzey
A poorly known species recognised by its scarcely exserted corolla 5–6 cm long with a yellow tube and purple limb, the lower lobes with darker margins, papillate greyish stigma; filaments glabrous, yellow. Host unknown. Native to Cameroon where it occurs at the foot of cliffs on seasonally flooded plains in peri-forest shrub savanna.

Aeginetia indica
in cultivation at
Oxford Botanic Garden, UK (right)

NAME ***Aphyllon arizonicum*** (L.T.Collins) A.C.Schneid.
SYNONYM ***Orobanche arizonica*** L.T.Collins

Habit: short, robust, spike-like inflorescences with flowers to ground-level. **Height**: 5–35 cm. **Stem**: base not swollen; inflorescence a dense raceme; scales acute, 0.5–1.1 cm, narrowly oblong-lanceolate. **Calyx**: rather long, lanceolate; deeply 5-lobed; lobes equal, densely glandular-pubescent; 0.8–1.2 cm; lobes 0.6–0.7 cm. **Corolla**: narrow tubular, sub-erect, later downward-curved; lobes acute but not apiculate, ciliate, exterior glandular-pubescent; 1.5–2.0 cm; tube externally white, apically dark purple; palatal folds yellow; exterior tube pubescent without glandular trichomes. Upper lip 0.4–0.6 cm, dark purple, divided into lobes to ½ the lip length, each lobe 0.2–0.3 cm across; lobes triangular-acute (rarely obtuse-rounded). Lower lip 0.3–0.4 cm, divided into 3 linear lobes to the base, each lobe 0.1–0.2 cm across, purple veined. **Carpel**: stigma somewhat 2-parted to entire, peltate, discoid-crateriform. **Stamens**: anthers glabrous or sparsely woolly from sutures of the thecae; hairs few. **Host**: Sonoran Desert shrubs in the family Asteraceae, especially *Gutierrezia sarothrae*. **Ecology and distribution**: Arizona, Nevada, New Mexico, Utah, in pinyon-juniper woodlands, deserts and on red sands. **Key diagnostics**: short, robust, purplish spikes with flowers to ground-level. **Similar species**: could be confused with *A. ludovicianum* and co-occurring with *A. multiflorum* and *A. corymbosum*. *Aphyllon ludovicianum* blooms in summer rather than late winter to spring. Variation in corolla lobe morphology (rounded, truncate, denticulate) can resemble *A. corymbosum*, but *A. multiflorum* and *A. corymbosum* have larger flowers, rounded corolla lobes and different hosts. Similar also to *A. riparium*, which differs in having a slightly flared, lavender or pallid white tube with dark purple veins, broadly triangular, purple or pale lavender upper lobe, and as it occurs in riparian habitats in association with different hosts.

Aphyllon arizonicum
Colorado, USA (left, right) photos: Craig Ontl

NAME ***Aphyllon californicum*** (Cham. & Schltdl.) A.Gray
SYNONYMS ***Orobanche californica*** Cham. & Schltdl.; ***Phelypaea californica*** (Cham. & Schltdl.) G.Don; ***Myzorrhiza californica*** Rydb.

Habit: low, clump-forming inflorescences; flowers racemose-corymbose, long-pedicellate, strong, resembling an oblong spike or cluster. **Height**: 5–15 cm. **Stem**: scarcely visible above-ground; stout, with numerous ovate basal scales to 1 cm long. **Calyx**: shortly bell-shaped, deeply 5-parted; teeth narrowly linear, arising from a wider base, shortly glandular-hairy, pinnately veined, equalling the corolla tube, surpassing top of spike. **Corolla**: slightly curved and constricted at the filament insertion point, and at the throat; 3.0–3.6 cm long; tube yellow to ochre below, fading to white, purple distally. Upper lip rather flat, bilobed, slightly curved upwards, elliptical lobes, rounded-acuminate, sometimes emarginate, very short glandular on both sides. Lower lip elongated or slightly bent, with equal lobes, elliptical or oblong, towards the apex acute or rounded narrow, sparsely glandular or glabrous; 3-lobed, lobes short and distant. **Carpel**: stigma funnel-shaped, often irregularly lobed, velvety, pale, whitish. **Stamens**: anthers tapering from a broad, rounded base, very shortly mucronate, with dense downy hairs to glabrous, depending on the subspecies. **Host**: parasitic on Asteraceae, especially genus *Grindelia*; also reported on *Erigeron glaucus*. **Ecology and distribution**: British Columbia, California, north-west Mexico, Oregon, Washington. Rare and possibly extinct in meadows of the Willamette Valley, Oregon; occasional on rocky banks and on sandy flats and slopes near the sea in south-west British Columbia and in north-west Washington on the San Juan Islands and occasionally on the neighbouring mainland; less often on sandy or gravelly substrates of coastal bluffs along the California coast. **Key diagnostics**: low, clump-forming yellowish, orange-red to purplish inflorescences borne at ground-level (appearing virtually stemless). **Similar species**: none.

Aphyllon californicum
(right) photo: Neal Kramer

Infraspecific taxa

Subsp. *condensum* has small stature, few flowers borne in a congested sub-corymbose or sub-capitate inflorescence, and deep purplish-red upper corolla lip lobes. Parasitic on *Chrysopsis villosa* on sandy or gravelly substrates, from 150 m to 500 m in California, from San Benito to Santa Barbara.

Subsp. *feudgei* has a rather broad corolla tube; co-occurs with its host *Artemisia tridentata* in the south-east of the species' range on washes and open sagebrush flats and slopes in California, from the Kern Plateau of Tulare Co. and the vicinity of Mt Pinos in Ventura Co. south through the south-western mountains to the Sierra San Pedro Mártir in Baja; 750–2,500 m.

Subsp. *grande* has buff to yellowish corollas with pinkish or pale brownish-red lips with darker veins; morphologically variable but differs from other subspecies in elevation, habitat and host preference. Parasitic on various genera including *Corethrogyne*, *Lessingia*, *Isocoma*, *Haplopappus*, *Adenostoma*, *Artemisia*, and *Heterotheca* in California, on dunes, bluffs and in coastal canyons from San Luis Obispo to Los Angeles and Santa Rosa Island.

Subsp. *grayanum* has dull white, yellowish, pinkish or pale lavender flowers, often with lavender veins. Parasitic on *Erigeron* and *Aster* in meadows and along stream margins, 1,200–1,800 m in California, descending to about 450 m in the Cascade Range of southern Washington; occasionally near sea level; also in the Cascade–Sierra Nevada ranges from Klickitat Co., Washington, south to Tuolumne Co., California; occasionally in Coast Ranges (south Oregon and central California) and mountains of the Columbia Plateau in Oregon.

Subsp. *jepsonii* has pale flowers – typically whitish, yellowish, brownish or pinkish with darker veins. Parasitic on *Grindelia*, *Baccharis viminea*, *Rubus vitifolius* and *Chrysothamnus* in lowlands and dry hillsides up to 3,000 m. Central Valley and surrounding mountains from the Cascade Range of Shasta and Plumas counties in the Sierra Nevada to Kern Co.; central California Coast Ranges from Tehama to Santa Cruz counties.

Aphyllon californicum
(right) photo: Neal Kramer

NAME ***Aphyllon cooperi*** A.Gray
SYNONYMS ***Orobanche cooperi*** (A.Gray) A.Heller; ***Myzorrhiza cooperi*** (A.Gray) Rydb.

Habit: densely clumped, purple spike-like inflorescences with flowers from ground-level. **Height**: 5–45 cm. **Stem**: simple or branched, swollen at the base, glabrous below; scales 0.5–1.3 cm, lanceolate to ovate, acute or obtuse; viscid-glandular-pubescent. **Calyx**: dark purple, deeply cleft into 5 lanceolate acute-attenuate lobes, densely glandular-pubescent; 0.7–1.2 cm. **Corolla**: an expanded, pubescent tube; corolla lobes with apiculate teeth; 1.5–3.2 cm long; exterior tube whitish, dark purple within; lobes each with 3 dark veins. Upper lip reflexed, erect to revolute, 0.6–1.0 cm triangular-acute with apiculate teeth. Lower lip 0.4–0.7 cm, lobes spreading or slightly reflexed. **Carpel**: stigma peltate, discoid, bilaminar, rhomboid, with entire to crenulate margins, white. **Stamens**: filaments glabrous; anthers glabrous or tomentulose along the sutures, sometimes with stalked glands on the dorsal surfaces. **Host**: Asteraceae: *Ambrosia*, *Sidneya* and *Encelia*. **Ecology and distribution**: deserts (notably the Sonoran and Chihuahuan), and other arid habitats in Arizona, California, central, north-east, north-west and south-west Mexico, Nevada, New Mexico and Texas. **Key diagnostics**: densely clumped grey-purple spikes in bare, arid habitats. **Similar species**: none.

Aphyllon cooperi
(right) photo: Michael Heine

NAME ***Aphyllon franciscanum*** (Achey) A.C.Schneid.
SYNONYM ***Orobanche fasciculata* var. *franciscana*** Achey

Habit: densely clustered (sub-corymbose), slender stems carrying sub-erect flowers. **Height**: 3–20 cm. **Stem**: single to many clustered, wiry stems; inflorescence a lax raceme, generally with 4–15 flowers; scales densely glandular-puberulent. **Calyx**: of similar colour to the corolla; 0.2–0.5 cm; lobes narrowly triangular to awl-like; the tips often darker. **Corolla**: narrowly tubular, sub-erect with rather long, slender, reflexed lobes; tube 1.5–3.2 cm; lobes 0.2–0.5 cm; pale to lemon yellow, sometimes tinged orange or purple. Upper and lower lip lobes narrowly ovate to obovate, deeply divided, divergent. **Carpel**: stigma 2-lobed, recurved, yellowish, with an orange nectary gland at the base of the ovary. **Stamens**: normally with hairy anthers. **Host**: parasitic on a range of desert plants, especially *Eriodictyon* and *Phacelia* (Boraginaceae), *Eriogonum* (Polygonaceae), also *Eriophyllum* (Asteraceae) and various Lamiaceae. **Ecology and distribution**: desert scrub in California, Oregon, British Columbia, Montana, Wyoming, Colorado, New Mexico, Texas and north-west Mexico. **Key diagnostics**: easily distinguished from other species by the striking yellow sub-corymbose inflorescences with sub-erect yellow flowers with pointed corolla lobes. **Similar species**: none.

Aphyllon franciscanum (above, right) photos: Michael Heine

NAME ***Aphyllon ludovicianum*** (Nutt.) A.Gray
SYNONYM ***Orobanche ludoviciana*** Nutt.

Habit: spike-like, purple racemes. **Height**: 7–40 cm. **Stem**: stout, simple or sparingly branched; purple, lavender or distally pale (rarely all yellow); scales 0.8–1.2 cm, lanceolate, densely glandular-pubescent. **Calyx**: deeply 5-lobed with lanceolate-subulate, glandular-pubescent lobes; 0.8–1.4 cm; purple, lavender or pale mauve. **Corolla**: tubular; 1.4–2.0 cm; white to pale cream, sometimes distally pinkish or purplish tinged and/or with purple veins. Lips externally white to pale cream, sometimes pinkish to purplish-tinged; internally pink or purple, sometimes white with purple veins, rarely light yellow. Upper lip with ovate, or sometimes deltate, lobes. Lower lip lobes oblong-lanceolate. **Carpel**: stigma pale yellow. **Stamens**: with filaments glabrous or pilose at the base; anthers glabrous or with a few hairs. **Host**: *Grindelia squarrosa* and *Artemisia* spp.; also reported on various other Asteraceae genera including *Baccharis*, *Haploesthes*, *Heterotheca* and *Thelesperma,* in southern Texas. **Ecology and distribution**: widespread from western Canada to Mexico in prairies, hillsides, dunes, bare ground, glades and roadsides. **Key diagnostics**: simple or sparingly branched, spike-like, purple racemes. **Similar species**: *Aphyllon riparium* is similar but has longer, less open flowers with lobes more acute.

Aphyllon ludovicianum Washington, USA (above, right) photos: Gerald D. Carr

NAME ***Aphyllon multiflorum*** (Nutt.) A.Gray
SYNONYM ***Orobanche multiflora*** Nutt.

Habit: simple or branched sparingly, relatively slender, the base scarcely enlarged; inflorescence sometimes thyrsoid, pale yellow, white, or tan proximally, purple or pale lavender distally, densely glandular-pubescent, appearing whitish or canescent, sometimes with axillary branches; flowers numerous. **Height**: 7–27 cm. **Stem**: scales numerous, appressed, lanceolate or broadly triangular, 3–10 mm, margins entire, sometimes ciliate, apex acute, surfaces glabrous. **Calyx**: 15–21 mm; pallid externally, purple internally, weakly bilaterally symmetric, deeply divided into 5 lobes, lobes lanceolate-attenuate, densely glandular-pubescent. Floral bracts erect to reflexed, ± lanceolate, 11–20 mm, apex acute or acuminate, densely glandular-pubescent. Pedicels 0–5 mm, much shorter than plant axis; bracteoles 2. **Corolla**: 22–36 mm; tube white to pallid or cream, sometimes pinkish or tinged light purple, rarely light yellow distally, sometimes with purple veins, constricted above the ovary, only slightly forward-inflected, densely pubescent; palatal folds prominent, yellow, densely pubescent; lips externally white to pallid or cream, sometimes pinkish or tinged light purple, internally pink or purple, sometimes white with purple veins, rarely light yellow, abaxial lip spreading, 5–9 mm, lobes broadly lanceolate, apex obtuse or rounded, adaxial lip erect or reflexed, 6–12 mm, lobes oblong, apex rounded. **Carpel**: stigma whitish or yellowish. **Stamens**: filaments pilose with a ring of hairs at the point of insertion; anthers included, woolly. **Host**: mainly on *Gutierrezia*, occasionally on *Heterotheca* and *Varilla* (Asteraceae). **Ecology and distribution**: Colorado, New Mexico and Texas in arid grasslands, semi-deserts and open woods at elevation 0–2,500 m. Flowering August–September. **Key diagnostics**: plants densely glandular-pubescent, appearing whitish-canescent, with racemes of numerous whitish or creamy, pink to purple-tinged flowers with deeply divided calyx lobes and pilose hairs around the filament insertion point; corolla lobes rounded; anthers woolly. **Similar species**: *Aphyllon ludovicianum* has a smaller corolla, the tube pale or whitish externally, with smaller lips; flowers rose or dark purple, anthers glabrous or pubescent, inflorescence uniformly glandular-pubescent (not whitish-canescent).

Aphyllon multiflorum Texas, USA (above, right) photos: Joey Santore

NAME ***Aphyllon parishii*** (Jeps.) A.C.Schneid.
SYNONYMS ***Orobanche californica* var. *parishii*** Jeps; ***Orobanche parishii*** (Jeps.) Heckard; ***Myzorrhiza parishii*** (Jeps.) Holub.

Habit: an erect, rather lax inflorescence. **Height**: 5–18 cm. **Stem**: solitary or clustered; inflorescence 3–8 cm, elongate, with flowers arranged in rather regular spirals along the axis. **Calyx**: concolourous with stem; lobes filiform, often twisted. 0.7–1.0 cm. **Corolla**: narrowly tubular-campanulate, conspicuously open, sub-erect, patent when mature; tube 1.5–2.4 cm long; lips 0.4–0.6 cm. Upper and lower lips strongly reflexed, with triangular-ovate, somewhat divergent lobes; buff to yellowish, often with conspicuous purplish-red, stripe-like veins along the interior lobes. **Carpel**: stigma lobes pale, narrow, recurved. **Stamens**: generally with glabrous anthers. **Host**: Asteraceae, normally *Isocoma menziesii*; also reported on *Haplopappus*. **Ecology and distribution**: California, including the Channel Islands, to Baja California in Mexico. Primarily on sandy maritime substrates near the ocean. **Key diagnostics**: flowers buff to yellowish with conspicuous stripe-like veins. **Similar species**: **subsp. *brachylobum*** is the form shown, distinguished from *A. parishii* subsp. *parishii* by its glabrous rather than hairy anthers.

Aphyllon parishii
(left, right) photos: Ron Vanderhoff

NAME ***Aphyllon pinorum*** (Geyer ex Hook.) A.Gray
SYNONYMS ***Orobanche pinorum*** Geyer ex Hook.; ***Myzorrhiza pinorum*** (Geyer ex Hook.) Rydb.

Habit: slender, somewhat lax inflorescence when mature, with regular, divergent branches (not solely from the base). **Height**: 10–30 cm. **Stem**: slender, swollen at the base, with many overlapping scales; deep purple, golden yellowish-brown or reddish; glandular-puberulent above-ground. **Calyx**: shortly pedicellate, with clasping narrowly triangular-acuminate lobes, slightly reflexed at the tips; 0.5–0.8 cm. **Corolla**: very narrowly tubular and strongly constricted at the throat; 1.2–2.0 cm. Upper and lower lips forward-pointing (not spreading or divergent); tube yellowish, with lobes concolourous tinged pale purple or brick red. **Carpel**: stigma 2-lobed, recurved, yellowish; filament bases with a hairy ring. **Stamens**: with glabrous or sparsely hairy anthers. **Host**: *Holodiscus* (Rosaceae). **Ecology and distribution**: widespread from south-west Canada to north-west California and New Mexico, on open, rocky forested slopes at elevations <2,100 m. **Key diagnostics**: slender inflorescence with regular, divergent branches and constricted corolla. **Similar species**: none.

Aphyllon pinorum Canada (above) photo: Erin Miller, New Mexico, USA (right) photo: Martin Purdy

NAME ***Aphyllon purpureum*** (A.Heller) Holub
SYNONYMS ***Orobanche porphyrantha*** Beck; ***Orobanche uniflora* var. *purpurea*** (A.Heller) Achey; ***Thalesia purpurea*** A.Heller

Habit: a small, solitary, 1–2-flowered inflorescence. **Stem**: short, 0.5–5 cm, rather thick, carrying 1–2 flowers. **Calyx**: conspicuously spreading with attenuate, thread-like lobes 0.4–0.8 cm, exceeding the width of the corolla tube. **Corolla**: smoothy arched, strongly constricted at the throat, inflated above, sub-erect, with a prominently open throat; 1.2–3.5 cm. Upper and lower lip lobes generally obtuse, minutely ciliate, weakly recurved; tube bicoloured yellow and purple; lobes whitish-purple with darker veins, yellow at the throat. **Carpel**: stigma 2-lobed with recurved margins, whitish, sometimes with a purple-blue tint; the ovary has an orange nectary gland at the base. **Stamens**: usually with hairy anthers. **Host**: parasitic on a range of herbs including *Senecio aronicoides* and *Antennaria* spp. (Asteraceae), *Sedum* (Crassulaceae), and various Apiaceae and Saxifragaceae. **Ecology and distribution**: California to British Columbia and the eastern Rocky Mountains; in moist habitats at elevations of <3,100 m. **Key diagnostics**: few flowers with yellow and purple colouration. **Similar species**: none.

Aphyllon purpureum USA (above, right) photos: Paul Hankamp

NAME ***Aphyllon riparium*** (L.T.Collins) A.C.Schneid.
SYNONYMS ***Orobanche riparia*** L.T.Collins; ***Myzorrhiza riparia*** (L.T.Collins) Weakley

Habit: solitary or clustered, lax inflorescences of terminal spike-like racemes, branched sparingly from the base. **Height**: 5–35 cm. **Stem**: stout or slender; mostly subterranean; glabrous; variably creamy-white, yellow, tan, ferruginous, dark brown or black; glandular-pubescent. **Calyx**: deeply 5-lobed; 0.7–1.1 cm. Lobes purple, linear-lanceolate to linear-subulate, densely glandular-pubescent. **Corolla**: tubular, constricted about the ovary; 1.5–2.2 cm long. Upper lip 0.5 cm with 2 triangular-acute lobes, lavender-purple, puberulent or pubescent. Lower lip 0.3–0.4 cm long with 3 linear lobes, pubescent, pale with 1–3 purple veins; tube cream-white sometimes tinged purple, the throat with yellow palatal folds. **Carpel**: stigma cream. **Stamens**: 4, generally glabrous or with few, scattered hairs. **Host**: reported mainly on the following Asteraceae: *Ambrosia trifida* and *Xanthium strumarium*, occasionally *A. artemisiifolia* and *Dicoria canescens*. **Ecology and distribution**: Colorado to West Virginia and Texas, primarily on mixed sand and silt deposits along the stream banks of eastern rivers, or sandbars of western rivers. **Similar species**: *Aphyllon ludovicianum* is similar but has a shorter corolla with flared throat and rounded lobes.

Aphyllon riparium Arizona, USA (above, right) photos: Doug Von Gausig

NAME ***Aphyllon robbinsii*** (Heckard) A.C.Schneid.
SYNONYM ***Orobanche robbinsii*** Heckard ex Colwell & Yatsk.

Habit: inflorescences composed of compact, branched corymbs. **Height**: 4–15 cm. **Stem**: simple or sparingly branched near the base, glandular-hairy. Aerial parts yellow, white, tan or infused with purple. **Calyx**: clasping, with linear, glandular lobes, 0.6–1.4 cm. **Corolla**: tubular-campanulate, narrow below, tapering into an abruptly open throat; 1.5–2.5 cm. Upper lip with 2 somewhat curved, reflexed, broadly triangular to oblong lobes, shallowly divided and divergent to form a v-like notch. Lower lip with 3 straight, strongly divergent, narrowly triangular to oblong lobes; tube pale creamy-white, tinged mauve, especially on the upper lip, with darker, simple veins along the lobes; throat with yellow folds. **Carpel**: stigma 2-lobed, irregularly pitted, creamy. **Stamens**: with filaments glabrous at the base. **Host**: *Eriophyllum staechadifolium* (Asteraceae). **Ecology and distribution**: south-west California on sandy or loose soils on coastal bluffs. **Key diagnostics**: compact, branched corymbs of flowers with spreading, oblong lobes, and restricted to one host, *E. staechadifolium.* **Similar species**: similar to *A. californicum* but distinguished by its glabrous anther sacs, its oblong, erect lobes and its host range.

Aphyllon robbinsii California, USA (above, right) photos: Dylan M. Neubauer

NAME ***Aphyllon tuberosum*** (A.Gray) A.Gray
SYNONYMS ***Orobanche bulbosa*** Beck; ***Myzorrhiza tuberosa*** (A.Gray) Rydb.; ***Orobanche tuberosa*** (A.Gray) A.Heller; ***Phelypaea tuberosa*** A.Gray

Habit: short, clumped, stout, broadly cylindrical to pyramid inflorescences of varying height, flowering from ground-level. **Height**: 8–30 cm. **Stem**: dark, dull blackish-purple with a whitish, bloom-like glandular-pubescent indumentum; thickened below-ground with overlapping scales. **Calyx**: concolourous with stem; tube dissected almost to the base beneath; lobes narrowly triangular, weakly reflexed at the tips. **Corolla**: narrowly tubular, constricted at the base and immediately below the throat, erecto-patent; throat rather closed; 1–1.8 cm long. Upper lip with 2 elliptical shallow curved upwards. Lower lip with 3 elliptical, scarcely reflexed to forward-pointing; tube dull yellowish- to blackish-purple throughout with many dense, short, whitish hairs. **Carpel**: stigma 2-lobed, recurved at the margins, usually whitish, sometimes with a shade of purple. **Stamens**: with glabrous to sparsely hairy anthers. **Host**: *Adenostoma fasciculatum* (Rosaceae). **Ecology and distribution**: dry, rocky habitats in California and north-west Mexico at elevations of <1,700 m. **Key diagnostics**: dull, blackish, congested spikes with narrowly tubular flowers with narrow throats. **Similar species**: none.

Aphyllon tuberosum
(left, right) photos: Paul Hankamp

NAME **Aphyllon uniflorum** (L.) A.Gray
SYNONYM **Orobanche uniflora** L.

Habit: simple or sparingly branched basally, sometimes forked medially; inflorescences of solitary flowers or fascicles of 2(–4) at stem or branch tips, white to cream, sometimes purple-tinged, simple, glabrous; bracts loosely ascending and erect, oblanceolate to broadly ovate, obovate, rhombic or awl-shaped, 5–12 mm, apex acute to acuminate, glabrous, rarely glandular-pubescent distally. **Height**: 3.5–18(–25) cm. **Stem**: slender, the base not swollen; scales relatively few, loosely imbricate to remote, loosely ascending to spreading; blade oblong-lanceolate to awl-shaped, 2–10 mm, margins entire, often incurved, apex acuminate, surfaces glabrous. **Calyx**: white to straw-coloured or light yellow, or pale to dark, dull purple, sometimes brownish, ± radially symmetric, 4–12(–15) mm, divided into 5 subequal lobes, lobes slightly shorter than to 2 × the length of the tube, ± triangular or subulate-triangular, sparsely to moderately glandular-pubescent. Pedicels (8–)20–110(–170) mm, much exceeding the plant's axis; bracteoles absent. **Corolla**: (11–)15–30(–35) mm, tube white to purple, blue or yellow, sometimes lighter proximally or with light or darker purple or blue veins, slightly to moderately constricted above the ovary, ± bent forward, glandular-pubescent; palatal folds ± prominent, bright yellow, glandular and/or eglandular-pubescent, sometimes glabrescent; lips white, yellow, purple or blue, sometimes with light or darker purple or blue veins; abaxial lip slightly to moderately spreading, sometimes ± recurved distally, (1–)2–6(–9) mm, lobes oblong-obovate to nearly round, apex rounded to bluntly pointed or shallowly emarginate, adaxial lip slightly to moderately spreading, sometimes recurved, 2–6(–9) mm, lobes broadly oblong-ovate to oblong-semi-orbiculate, apex rounded, rarely bluntly pointed or shallowly emarginate. **Carpel**: stigma usually whitish. **Stamens**: filaments glabrous; anthers glabrous or villous-tomentose. **Host**: parasitic on a range of genera including *Sedum*, *Saxifraga*, *Helianthus*, and *Solidago* in Alaska. **Ecology and distribution**: anthropogenic habitats, forests, meadows and fields, woods in the eastern United States, and southern Canada. Flowering April–August. **Key diagnostics**: a cool-temperate, few-flowered species with 1–5 flowering stalks, each with a single flower; pedicels are longer than the stems; calyx segments awl-shaped with long tips and the lobes longer than their tube; corolla white to lilac with bright yellow or creamy-white hairs along the lower throat. **Similar species**: *A. fasciculatum* (polymorphic complex) differs, usually having branched forms, stouter stems with more flowering pedicels (3–12), stems and pedicels about equal in length; calyx lobes are triangular and shorter than, or about equalling, the tube.

Aphyllon uniflorum
Connecticut, USA
(left, right) photos: Eric Powell

Additional species

NAME ***Aphyllon fasciculatum*** (Nutt.) Torr. & A.Gray

Forms a polymorphic complex, similar to *A. uniflorum*, that may include cryptic species, but differs in having stouter stems with more flowering pedicels (3–12), stems and pedicels about equal in length; calyx lobes triangular and shorter than, or about equalling, the tube. Found on a broad range of hosts, mainly *Artemisia* (Asteraceae), *Phacelia* (Boraginaceae), *Eriodictyon* (Boraginaceae) and *Eriogonum* (Polygonaceae). Flowering April–August. Mainly in upland prairies, on dunes, desert scrub, rocky slopes, forests, alpine meadows, at elevation 150–3,300 m, from Alaska to Mexico (Baja California Norte, Chihuahua).

NAME ***Aphyllon tacnaense*** (Mattf.) A.C.Schneid.

A subtropical species recognised by its violet flowers with rounded lobes, bilobed stigma, and glabrous filaments inserted above the corolla base. Parasitic on *Artemisia* in Peru. Easily confused with *A. tuberosum* but has an extended, slenderer (less squat) inflorescence and larger flowers, with broader corollas.

NAME ***Aphyllon tarapacanum*** (Phil.) A.C.Schneid.

Recognised by its dense clusters of dark yellow flowers with brown-violet lobes, and a stripe of colour along each lobe extending into the tube; upper lip with 2 rounded lobes; filaments glabrous; calyx with 15 veins. Only reported to be parasitic on *Trixis cacalioides* in the Tarapacá region of northern Chile in the Andean Desert, coastal deserts and semi-deserts.

NAME ***Aphyllon validum*** (Jeps.) A.C.Schneid.

A Californian species recognised by its dark brownish-purple spikes of flowers; lobes with whitish margins and dark mid-veins. Parasitic on the roots of trees and shrubs in southern California. **Subsp. *validum*** has puberulent corolla exterior, glabrous anthers and filaments glabrous at the base; parasitic on *Garrya veatchii* and *Eriodictyon trichocalyx.* **Subsp. *howellii*** is distinguished by its short-pilose corolla exterior, pilose anthers, and sparsely pilose filament bases; parasitic on *Garrya fremontii* and *Quercus chrysolepis* on volcanic and serpentine slopes in open chaparral, at elevations 200–1,700 m.

NAME ***Aphyllon vallicola*** (Jeps.) A.C.Schneid.

A Californian endemic recognised by its yellow, rose-tinted stems with pale yellow, pink-veined flowers. Parasitic on *Sambucus*; also reported on *Pyrus*, *Quercus agrifolia*, *Symphoricarpos albus* and *Baccharis douglasii.* Similar to *A. californicum* subsp. *jepsonii* but with crateriform rather than laminate stigma lobes.

NAME ***Aphyllon weberbaueri*** (Mattf.) A.C.Schneid.

A poorly known subtropical species native to Peru with ferruginous indumentum and white-yellow flowers, often with purple lips; lobes broadly triangular. Host unknown.

Aphyllon fasciculatum
(right) photo: Paul Hankamp

NAME ***Boschniakia rossica*** (Cham.& Schltdl.) B.Fedtsch.

SYNONYMS ***Boschniakia glabra*** C.A.Mey. ex Bongard; ***Orobanche glabra*** Hook.; ***Orobanche rossica*** Cham. & Schltdl.

Habit: robust, simple, spike-like inflorescences with flowers almost to ground-level and with numerous scales around the stem base. **Height**: 15–35 cm. **Stem**: solitary or few (1–3), erect, dark wine red or purple, unbranched, with rhizome-like, cylindrical to globular base below-ground, 1–2.5(–4) cm across. Inflorescence spicate, 7–22 cm, 1.5–2.5 cm in diameter. Bract glabrous abaxially, pubescent adaxially, bracteoles absent. **Calyx**: purple or red-brown; lower floral bracts conspicuous: broad, pale and adpressed to the corolla base. **Corolla**: short-tubular, glabrous except the margins; tube broadly campanulate, the tube base ± inflated, with an open throat with style and stamens often partially protruding; 8 × 10 mm. Upper lip 5–7 mm, lobes ca. 1 mm. Lower lip very short, 3-lobed; corolla dark purple-red throughout, with conspicuous white-ciliate margins. **Carpel**: stigma 2-lobed, whitish or yellowish; ovary ovoid; style base persistent in fruit, 5–7 mm, glabrous. **Stamens**: included or exserted; filaments 5.5–6.5 mm; anthers ovoid, ca. 1.2 mm, glabrous. **Host**: *Alnus* spp. (Betulaceae). **Ecology and distribution**: distributed from north-east Russia through temperate Asia to Alaska and western Canada, in temperate forests, bogs, both coastally and in the interior along streams and lakeshores and mountain slopes 0–1,800 m. Flowers May–July (August) depending on distribution and altitude. **Key diagnostics**: unmistakable for its robust, spike-like, wine-red inflorescences. **Similar species**: *Kopsiopsis* spp. have compact inflorescences or open racemes, pedicels and bracteoles present, a more funnelform corolla, and grow on Ericaceae; somewhat similar to *Xylanche himalaica*, which differs in having solitary, sombre-coloured spikes with hooded, open-mouthed flowers, as well as a different distribution and host range.

Boschniakia rossica Japan (above, right, pages 96, 97) photos: Kenji Suetsugu

NAME ***Christisonia bicolor*** Gardner
SYNONYMS ***Christisonia aurantiaca*** Wight; ***Cliffordia bicolor*** (Gardner) Livera

Habit: short, robust, clumped; flowers and the flower clusters stalked. **Height**: 8–12 cm. **Stem**: thick, pubescent, brownish-yellow with few scales. **Calyx**: 1.8 cm long, reddish-orange, brick-red, tube broad, villous to sub-glabrous; calyx-teeth acute, lobes short, triangular, long-pointed, glandular-hairy. Bracteoles 2.8 × 3 mm, lanceolate, pointed, velvety. **Corolla**: tube narrow below, broadened above; 6 cm long, with a hairy exterior, bright to deep yellow, with rather indistinct upper and lower lips; lobes orbicular, glandular-hairy. **Carpel**: stigma peltate with a large pit at centre, whitish to yellowish; style to 4.5 cm long. **Stamens**: filaments 2.5 cm long, glandular-hairy throughout; anther lobes 3 mm long, spur-like, acuminate. **Host**: *Strobilanthes* (Acanthaceae). **Ecology and distribution**: wet tropical forests in southern India and Sri Lanka. **Key diagnostics**: a conspicuous bright yellow and reddish-orange plant, often patch-forming. **Similar species**: none.

Christisonia bicolor India (above, pages 100, 101) photos: Siddarth Machado, Sri Lanka (right) photo: Vinolperera

NAME ***Christisonia calcarata*** Wight
SYNONYMS ***Christisonia lawii*** Wight; ***Christisonia stocksii*** Hook.

Habit: short, robust, clumped, with erect, crowded flowers. **Height**: 10–20 cm. **Stem**: slender, 2–3.5 cm long, covered with whorled, ovoid-elliptic bracts (the uprooted stem resembles a bromeliad). Flowers and buds yellow-purple, raised on glabrous pedicels 2–10 cm long; buds covered with copious, translucent, tasteless slime when young. **Calyx**: glabrous, yellow, tubular, 1.5–2.5 cm, 5-toothed; teeth 2–6 mm long, ovate, acute, cuspidate. **Corolla**: 4–6 cm long, tubular-campanulate, somewhat curved; corolla lobes 5, equal to unequal; tube covered with white pubescence outside and densely pubescent at the base, bright yellow and glabrous inside; lobes rounded to acute at the apex, with undulate margins, creamy whitish-blue, 1–2 cm long; lower lips smaller than the uppermost. **Carpel**: stigma bilobed, with 2 ovoid, folded discs. **Stamens**: 4, included, didynamous, the lower pair ½ the length of the upper pair; sparsely scattered with short hairs on the filaments; filaments white-pubescent; anthers pale to white, 2-celled. **Host**: *Strobilanthes* and *Pleocaulis* (Acanthaceae). **Ecology and distribution**: humid forests on high to medium elevation slopes and plateaus in Pakistan to western India and Sri Lanka. Flowering July–September. **Key diagnostics**: small plant with a whitish corolla, often clumped; corolla tube exterior and both sides of the lobes white-pubescent, stigma appearing bilobed in lateral view. **Similar species**: *Christisonia scortechinii* and *C. hookeri* differ in having a glabrous corolla, glabrous filaments, whitish to violet (not yellow) calyx and white to violet-tinged corolla.

Christisonia calcarata India (above) photo: Mayur Nandikar, (right) photo: Prajwal J. Ullal

NAME ***Christisonia hookeri*** C.B.Clarke ex Hook.f.

Habit: short, robust, clumped; flowers 2 to several clustered at the stem apex. **Height**: 3–8(–12) cm. **Stem**: 1–2 cm, sub-glabrous; scales few, ovate. **Calyx**: tubular, 1.5–3 cm, sub-membranous or leathery when dry, apex irregularly (4)5-lobed; lobes triangular or lanceolate, 2 larger, 0.5–1 cm, 3 smaller, 4–8 mm; apex usually acute; white with violet seams, glabrous. Pedicel short and thick, ca. 1 cm long, or absent. Floral bracts oblong or ovate, 8–10 × 5–6 mm. **Corolla**: white or rarely pale purple, tubular, 2–7 cm; lobes 5, subrounded, entire, corolla tube interior yellow. **Carpel**: stigma whitish, prominent and bilobed (in some reports discoid); style 2.5–4 cm. **Stamens**: filaments 8–10 mm, glabrous or sparsely glandular pubescent at the base; anthers with 1 fertile cell in upper 2 stamens, the lower 2 reduced; anthers glabrous. **Host**: unknown. **Ecology and distribution**: east Himalaya, montane humid forests in India, Laos, Sri Lanka, Thailand, at elevation 1,500–2,000 m. **Key diagnostics**: a small plant with whitish corolla, often clumped; corolla glabrous; stigma appearing bilobed in lateral view. **Similar species**: other whitish species such as *C. scortechinii, C. calcarata* and *C. sinensis* (also considered a synonym of *C. hookeri*) differ in stigma morphology: bilobed in *C. hookeri* and *C. calcarata*, and peltate in the other species. *Christisonia calcarata* also has a distinctive white-pubescent corolla, hairy anthers and yellow calyx.

Christisonia hookeri
Taiwan (right)
photo: Zhi-Xiang

NAME ***Christisonia scortechinii*** Prain
SYNONYM ***Aeginetia scortechinii*** (Prain) Livera

Habit: low-growing on the forest floor with few, glabrous flowers, often partially obscured by leaf fall. **Height**: ground-level. **Stem**: above-ground stems absent; pedicels ca. 1 cm long. **Calyx**: white to purplish, glabrous with 2–4(–5) lobes. **Corolla**: sub-erect, broadly tubular, open-throated with short, blunt triangular lobes, the limb oblique in profile; corolla pure white with a conspicuous yellow to dark orange spot in the throat; flowers rarely with purple concolourous around the lobe margins. **Carpel**: stigma peltate, white, inserted in the corolla tube. **Stamens**: glabrous. **Host**: parasitic on bamboos such as *Racemobambos gibbsiae* (Poaceae). **Ecology and distribution**: humus-rich substrates on the forest floor in montane sites at elevations of ca. 650–1,700 m, in Peninsular Malaysia, Thailand, Cambodia and Vietnam (probably also in the Philippines). Flowering year-round. **Key diagnostics**: glabrous white flowers with a conspicuous dark orange spot; low-growing habit and tropical ecology. **Similar species**: purple-tinted colour morphs can resemble *C. siamensis* but *C. scortechinii* always has glabrous stamens.

Christisonia scortechinii Cambodia (above, right) photos: Oleg Kosterin

NAME ***Christisonia siamensis*** **Craib**
SYNONYM ***Aeginetia siamensis*** **(Craib) Livera**

Habit: low-growing on the forest floor with few, glabrous flowers, often partially obscured by leaf fall; buds covered in slime when immature. **Height**: 10–13 cm. **Stem**: above-ground stems absent; pedicels ca. 1 cm long. **Calyx**: white, yellowish to dark purple, glabrous, with (2–)3(–4) deltoid teeth. **Corolla**: tube white to dirty white, strongly pigmented with violet distally and around the lobe margins; throat with a yellow to dark orange stripe. **Carpel**: stigma peltate, white, inserted in the corolla tube. **Stamens**: a distinct tuft of hairs at the point of filament insertion that obstructs the corolla throat; filaments yellow. **Host**: reported to be parasitic on *Tetrastigma* (Vitaceae); host range requires verification. **Ecology and distribution**: mixed deciduous forests in Thailand, Nagaland, north-east India and south-west China at elevations of 300–800 m. **Key diagnostics**: violet corolla colouration and a distinct tuft of hairs at the point of filament insertion that obstructs the corolla throat make this species easily recognisable. **Similar species**: purple-tinted colour morphs of *C. scortechinii* can be somewhat similar but this species typically lacks the strong purple colouration of *C. siamensis*, and *C. scortechinii* always has glabrous filaments.

Christisonia siamensis
Thailand (right)
photo: Muangpai Suetrong

NAME ***Christisonia tubulosa*** (Wight) Benth.ex Hook.f.
SYNONYM ***Oligopholis tubulosa*** Wight

Habit: lax, few-flowered. **Height**: 15–26 cm. **Stem**: rather slender and flexuous, with obtuse scales 2–3 mm long, very few, distant, acute, ovate-lanceolate. Stem simple or branched at the base. **Calyx**: tubular, to 1.2 cm long, with small, triangular ± acute lobes. **Corolla**: rather large and conspicuous relative to the stems; tube narrowly funnel-shaped, campanulate and abruptly flared at the mouth; 4.0–6.5 cm long. Upper and lower lip with 5 rounded, subequal, entire lobes; floral tube externally whitish, strongly tinted reddish-purple or violet-blue at the apex; yolk-yellow in the throat. Pedicels 2.5–10 cm long. **Carpel**: stigma orbicular, white, large and conspicuous, semi-occluding the throat. **Stamens**: anther-spurs acute; filaments pubescent, didynamous, in pairs; anthers 2-celled, 1 cell perfect, other sterile, spurred. **Host**: unknown. **Ecology and distribution**: seasonally dry tropical forests of the southern and western Ghats in India, between 900 m and 1,200 m. Flowering and fruiting reportedly in July–August. **Key diagnostics**: few, large, whitish, strongly tinted reddish-purple or violet-blue flowers with yellow throats, in seasonally dry tropical forests. **Similar species**: none.

Additional species

NAME ***Christisonia albida*** Thwaites ex Hook.f.
A distinctive, entirely white species with crowded inflorescences with very short pedicels (1.2–1.3 cm) and rather large flowers to 7 cm; stamens inserted; calyx glabrous. Host unknown. Native to the seasonally dry tropical forests of Sri Lanka. Similar to *C. bicolor* but entirely white and with a glabrous calyx.

NAME ***Christisonia flammea*** Sedgw.
A golden-yellow plant with erect stems and flowers borne in crowded and corymbiform inflorescences. Corolla yellow, trumpet-shaped, the tube inflated at the middle, the limb with 5 equal, very obtuse lobes; exterior densely pubescent; calyx reddish-brown. Stigma 2-lobed. Parasitic on *Strobilanthes* in seasonally dry tropical forests in India.

Christisonia tubulosa
India (left, page 112)
photos: Srinivasan Kasinathan
Illustrated (right)
India (page 113) photo: Ambady Sasi

NAME ***Cistanche armena*** (K.Koch) M.V.Agab.
SYNONYM ***Phelypaea armena*** K.Koch

Habit: unbranched, erect, elongated and rather lax inflorescences. **Height**: 22–70 cm. **Stem**: unbranched, glabrous below, covered with fleshy, ovate, acute scales, those above white-woolly; yellow to brown or violet-coloured, swollen at the base. **Calyx**: ovate, acute, with unequal lobes; ½ as long as the bracts and ⅓ as long as the corolla. Floral bracts ovate-lanceolate, with linear-lanceolate bracteoles. **Corolla**: tube strongly curved for most of its length, lobes suborbicular, as long as wide, and remaining somewhat concave; tube 22–30 mm, basally white, distally becoming reddish-violet. Upper lip with 5 violet, suborbicular lobes. **Carpel**: stigma is sub-globose, white. **Stamens**: filaments hairy, obliquely inserted 4–6 mm above the base of corolla, with long hairs at the base and short glandular hairs above, anthers densely villous, 3.5–4 mm. **Host**: Camelthorn (*Alhagi maurorum*, Fabaceae), saltwort (*Salsola dendroides*, Amaranthaceae) and *Halostachys belangeriana* (Amaranthaceae). **Ecology and distribution**: endemic to Armenia. Flowering May–early June and fruiting June–July. It is known only from the Ararat and Armavir provinces in central Armenia, at elevation 820–840 m, in arid areas, semi-deserts, near irrigation ditches, on sandy and saline substrates. **Key diagnostics**: pinkish-mauve, elongated, lax inflorescences; flowers strongly curved. **Similar species**: *Cistanche salsa* is a more compact plant with flowers that hide the inflorescence axis, and the corolla tube is completely white, curved outward only near its tip.

Cistanche armena
Armenia (left, right)

NAME *Cistanche deserticola* Ma

Habit: very robust, unbranched, erect, dense and many-flowered spike. **Height**: 0.4–1.6 m. **Stem**: robust, 2–10 cm in diameter at the base; extensive underground; fleshy and scaly; scales linear-lanceolate to lanceolate; those below ovate to triangular-ovate, 0.5–1.5 × 1–2 cm; those above lanceolate to linear-lanceolate, 2–4 cm × 5–10 mm, pubescent. **Calyx**: campanulate; 1–1.5 cm; lobes 5, ca. 2.5 × 3–5 mm (subequal to corolla). Floral bracts brown, narrow and attenuate, recurved, subequalling the flowers (exceeding the buds). **Corolla**: tubular-campanulate; 3–4 cm, apex 5-lobed; lobes 4–6 × 6–10 mm; variable in colour, cream or white, variably flushed with purple and with yellow folds in the throat; fragrant. **Carpel**: stigma sub-globose; style glabrous, usually persistent. **Stamens**: filaments 1.5–2.5 cm, villous at the base; anthers long, ovoid, 3.5–4.5 mm, densely villous, basally mucronate. **Host**: Predominantly *Haloxylon ammodendron* and *H. persicum* (Amaranthaceae); also other desert shrubs and cultivated on *Atriplex*. **Ecology and distribution**: China: Gansu, Inner Mongolia, Ningxia, Xinjiang; also known from Mongolia, in sandy desert areas dominated by halophytic shrubs at 200–1,200 m. Rare and endangered in the wild but now cultivated extensively. Flowering May–June depending on locality, rainfall and altitude (generally fruiting July–August). **Key diagnostics**: long, dense, pale inflorescences with numerous congested flowers; pubescent; desertic ecology. **Similar species**: *Cistanche salsa* (much-confused with *C. deserticola* in China) is a smaller plant and has ovate-lanceolate bracts ½ as long as the corolla; the plant is morphologically very similar to some forms of *C. deserticola,* but does not occur as far east as China, as has been confirmed by molecular data.

Cistanche deserticola Inner Mongolia, China (above, right, page 119), China (page 118) photo: Majed Aldughayman

NAME ***Cistanche fissa*** (C.A.Mey.) Beck
SYNONYMS ***Orobanche salsa* f. *fissa*** (C.A.Mey.) Kuntze; ***Phelypaea fissa*** C.A.Mey.

Habit: rather short, squat. **Height**: to 30 cm. **Stem**: pale, whitish to pale purple; scales subacute, 15–16 × 3.5–4 mm, glabrous, entire, non-scarious at the margins. **Calyx**: short-campanulate, about ½ as long as the corolla, arachnoid-woolly on the outer side and densely woolly at the margins, lobes 5, unequal. Floral bracts ovate to oblong-lanceolate, subacute, 16–17 × 5–5.5 mm, densely woolly on the outer side, longer than the calyx, entire and non-scarious; bracteoles oblong-linear, acute, as long as calyx or slightly longer, woolly at their margins and slightly woolly on the outer side. **Corolla**: tubular, 28–30 mm long, cream to pale purple with 2 yellow folds at the throat, interior glabrous; lobes 5, equal, rounded, pale pink to purple, densely woolly inside and at the margins. **Carpel**: stigma pale, whitish; inserted. **Stamens**: inserted near the base of the tube, woolly. **Host**: shrubby Amaranthaceae (*Salsola* spp., *Caroxylon* spp.) and possibly Asteraceae. **Ecology and distribution**: deserts, semi-deserts, steppe, slopes, roadsides, near cultivated fields, irrigation ditches. Flowering usually April–mid-May. Distribution improperly known; from western and central Asia to the Middle East (Afghanistan, Azerbaijan (type location), Georgia, Armenia, Iran, Israel, Turkmenistan). **Key diagnostics**: arachnoid-woolly indumentum; calyx lobes very unequal; corolla lobes abaxially woolly, corolla tube white, lobes light violet. **Similar species**: often confused with *C. salsa* which has equal to subequal calyx lobes, white-woolly (rather than arachnoid) indumentum in bud, and sub-glabrous corolla limb. ***Cistanche mimii*** Ben-Natan & Thorogood, recently described from Israel and Jordan, has similar calyx morphology to *C. fissa* but a floccose-lanuginose indumentum and cream, red-mauve tinted corolla with internal dark spots; fragrant.

Cistanche mimii Israel (left)
photo: Dar Ben-Natan
Cistanche fissa Azerbaijan
(right, page 122, 123)

NAME ***Cistanche flava*** (C.A.Mey.) Korsh.
SYNONYMS ***Cistanche tubulosa* var. *flava*** (C.A.Mey.) Beck; ***Phelypaea flava*** C.A.Mey.

Habit: robust, many-flowered spikes. **Height**: up to 1 m. **Stem**: robust, with numerous, conspicuous, long-lanceolate, obtuse scales, 40–85 mm long, glabrous. **Calyx**: tubular-campanulate with 5 lobes, more deeply incised between 2 lobes than between the remining 3; lobes unequal, glabrous. Floral bracts oblong-lanceolate, serrate, twice as long as the calyx, grey to purple; floral bracteoles oblong-linear, obtuse, slightly serrate at the margins, entire, just shorter than the calyx, grey to purple, glabrous. **Corolla**: campanulate, 37–60 mm long, yellow-white outside with darker veins inside; lobes 5, yellow or blue-violet, 5–9 × (4–)7–10 mm. **Carpel**: stigma white or yellowish. **Stamens**: filaments inserted near the corolla base, woolly. **Host**: *Tamarix* spp. (Tamaricaceae), possibly also on shrubby Amaranthaceae; reported also on *Ephedra* (Ephedraceae). **Ecology and distribution**: semi-deserts, deserts, tamarisk groves along watercourses and rivers, on coastal dunes, and in halophyte communities in south-west and central Asia; Azerbaijan is the type location. Flowering usually May–June. **Key diagnostics**: stem scales and floral bracts typically long (scales shorter in some populations); floral bracts at least twice as long as calyx. **Similar species**: the widespread *C. laxiflora* (type location: Afghanistan) appears to be virtually indistinguishable from *C. flava*; the distinction between these taxa is unclear. Plants with very dense, congested spikes of bright violet and yellow flowers in the deserts of Kazakhstan also require further investigation.

Cistanche flava Azerbaijan (above, right, page 126, 127)

NAME ***Cistanche mauritanica*** (Coss. & Durieu) Beck
SYNONYM ***Phelypaea mauritanica*** Coss. & Durieu

Habit: small, squat, congested spikes; conspicuously hairy in bud. **Height**: 40 cm. **Stem**: with scales densely arachnoid-hairy. **Calyx**: campanulate, 13–16(–20) mm, conspicuously arachnoid-woolly on the outer surface. Floral bracts ovate, acute, equal to or slightly shorter than the calyx, with long white hairs; bracteoles oblong-linear, as long as calyx or slightly shorter than the calyx. **Corolla**: arachnoid-hairy indumentum; tubular, 33–38 mm long, pale purple, with 2 yellow folds inside the tube; lobes unequal. **Carpel**: stigma pale. **Stamens**: filaments inserted near the tube base, woolly. **Host**: shrubby halophytes in the Amaranthaceae (*Atriplex*, *Salsola*, *Suaeda*) and *Lycium* (Solanaceae). **Ecology and distribution**: Islas Chafarinas off the coast of Morocco, and Algeria: a rare endemic of the north-west littoral of Africa. **Key diagnostics**: small habit, bracts, bracteoles and calyx densely covered with long arachnoid-hairy indumentum; bract usually as long as wide. **Similar species**: possibly confused with *C. violacea*, which is typically more robust, glabrous and unmistakable based on the combination of characters stressed above.

Cistanche mauritanica
Algeria (left) photo: Abdenour Kheloufi
Morocco (right, page 130, 131)
photos: Mimoun Bachiri

NAME ***Cistanche phelypaea*** (L.) Cout.

SYNONYMS ***Cistanche brunneri*** (Webb) Bég.; ***Cistanche hesperugo*** (Webb) Beck; ***Cistanche lusitanica*** J.A.Guim.; ***Lathraea phelypaea*** L.; ***Orobanche phelypaea*** (L.) Wallr.

Habit: robust, unbranched inflorescences, which are lax (Portugal) to somewhat congested (western Africa and Canary Islands); stems often devoid of flowers below. **Height**: up to 45 cm, sometimes there are also taller specimens. **Stem**: swollen at the base, yellow to grey-violet (or even blackish) and glabrous with robust scales, 1.8–2.5 × 0.3–0.8 cm, which are ovate-lanceolate with denticulate margins. Inflorescence variably dense and 5–9 cm wide and 10–20 cm long; scales 1.5 × 0.3–0.4 cm, lanceolate. **Calyx**: lobes overlapping, broadly ovate; calyx tube 12–18 mm, lobes 3–4 mm. Bracts ovate and ± equalling the calyx. Bracteoles shorter than the calyx, 1–1.2 × 0.2 cm, narrowly lanceolate. **Corolla**: tube sub-erect, distally curved with 5 broadly rounded lobes; tube 30–50 mm; typically uniform bright yellow (occasionally pale), although lobes may have mauve edges (especially plants in the south). **Carpel**: stigma pale yellow. **Stamens**: anthers densely hairy; filaments sparsely hairy at the base. **Hosts**: numerous coastal shrubs, especially Amaranthaceae including: *Arthrocnemum macrostachyum*, *Sarcocornia fruticosa* and *Suaeda vera*. Also reported (usually without verification) on species from other families such as *Limoniastrum monopetalum* (Plumbaginaceae), *Salvadora persica* (Salvadoraceae), *Bridelia somalensis* (Phyllanthaceae) and *Boerhavia* (Nyctaginaceae). **Ecology and distribution**: an Atlantic species that occurs from Portugal, south to coastal Morocco and Cape Verde (records elsewhere, for example in the Middle East pertain to other entities, especially *C. tubulosa*); typically on coastal dunes, alluvial sands and estuaries; often intertidal (periodically submerged). Visited by hawkmoths. Flowering December (south of range) to May (north of range). **Key diagnostics**: robust yellow spikes in saline habitats. **Similar species**: yellow-flowered species throughout the range of *Cistanche*: morphology alone is unreliable for teasing apart cryptic taxa. Populations in southern Spain often described as *C. phelypaea,* or forms thereof, appear to correspond better with North African entities (see notes on *C. lutea* under *C. violacea*). *Cistanche phelypaea* appears to be best defined by its ecology, as an exclusively coastal species. We have observed a faint fragrance in specimens in the Canary Islands.

Cistanche phelypaea
Fuerteventura, Canary Islands (left, page 135)
Portugal (right)
Lanzarote, Canary Islands (page 134)

NAME ***Cistanche rosea*** **Baker ex Hook.**

Habit: robust, unbranched, erect spike. **Height**: 30–45 cm. **Stem**: fleshy, slightly bulbous at the base, densely scaly with overlapping obtuse, ovate or ovate-oblong scales becoming narrower above. **Calyx**: a short, ovoid-elliptic tube extending into obtuse lobes; floral bracts oblanceolate or ovate-oblong, shorter than the flowers; bracteoles narrowly spathulate. **Corolla**: a slightly curved, funnel-like tube extending into 5 rounded lobes with blunt edges; tube 5 cm, rose-pink outside, paler within. Upper lip bilobed with rounded lobes. Lower lip with 3 broadly rounded lobes, the middle the largest. **Carpel**: stigma protruding, yellowish; ovary lanceolate-oblong with an elongated, glabrous style. **Stamens**: protruding; filaments villous at the base, inserted at the middle of the corolla tube; anthers obtuse, oblong, white and woolly. **Host**: *Pluchea dioscoridis* (Asteraceae). **Ecology and distribution**: deserts and dry scrub of Oman, Saudi Arabia and Yemen. **Key diagnostics**: pink colouration. **Similar species**: no other *Cistanche* species shares this colouration.

Cistanche rosea Oman (above) photo: Joey Bom, (right) photo: Youp van den Heuvel

NAME ***Cistanche salsa*** (C.A.Mey.) Beck

SYNONYMS ***Cistanche ambigua*** (Bunge) Beck; ***Cistanche eremodoxa*** Bornm.; ***Orobanche salsa*** (C.A.Mey.) Kuntze; ***Phelypaea salsa*** C.A.Mey.

Habit: robust, unbranched, rather squat, erect and very congested spike. **Height**: 10–40 cm. **Stem**: 5 cm thick and bulbous at the base, covered with overlapping obtuse ovate-oblong scales, the uppermost oblong-lanceolate, acuminate. **Calyx**: campanulate, as long as are wide, sometimes sparsely hairy but often glabrous. Lobes 5, ovate or semi-orbicular, glabrous or downy on the margins, densely veined, 2.5–3 mm × 3–3.5 mm; 5th lobe sometimes smaller; pale yellow or white; ½ length of corolla. Floral bracts tomentose, ovate-lanceolate, obtuse or acuminate, particularly downy on the margin; more sparsely abaxially; ⅓ the length of the corolla (1–1.5 cm). Bracteoles linear-oblong, of similar length to the calyx. **Corolla**: slightly curved, tubular-campanulate, glabrous, with 5 subrounded lobes; 2.5–2.7 cm long; each lobe 5–7 mm × 5–7 mm; white, purple towards the edges. Upper lip: with subequal semi-orbicular lobes, crenulate-denticulate along the margins. Lower lip: lobes similar, recurved. **Carpel**: stigma sub-globose, emarginate; style: glabrous, 1.6–2 cm, persistent. **Stamens**: filaments 12–14 mm, villous at the base; anthers long, ovoid, densely woolly, ca. 2.5 mm. **Host**: numerous shrubs from the Amaranthaceae, Tamaricaceae, Nitrariaceae and/or Asteraceae reported, especially: *Haloxylon ammodendron*, *Calligonum polygonoides*, *Tamarix aphylla*, *Kalidium foliatum*, *K. gracile*, *K. cuspidatum*, *Reaumuria songarica*, *Nitraria sibirica*, *Achnatherum splendens*, *Suaeda physophora*, *Krascheninnikovia ceratoides*, *Sympegma regelii* and *Salsola passerina*. **Ecology and distribution**: recorded in European Russia and Turkey, west to central Asia and east to China but much confused with other taxa (records in China correspond with *C. deserticola*). **Key diagnostics**: short, robust, purplish with hairy-edged bracts; violet-scented. **Similar species**: *Cistanche deserticola* typically has much larger, many-flowered inflorescences and bracts that equal the corolla. Polyphyletic in its current circumscription; requires further work to distinguish entities from Kazakhstan (the type locality) from what may be distinct taxa elsewhere.

Cistanche salsa
(right) photo: Vladimir Kolbintsev

NAME *Cistanche sinensis* Beck

Habit: unbranched, rather squat, erect spike. **Height**: 15–70 cm. **Stem**: 4–10 mm across and swollen at base, with abundant yellow glandular hairs; scales ovate-triangular to ovate-lanceolate, 5–20 × 4–8 mm. **Calyx**: campanulate, 12–22 mm, 4-lobed, each lobe linear or oblong-lanceolate, 10–12 mm, acuminate at the apex; floral bracts ovate-lanceolate to linear-lanceolate, 1.6–2 cm × 3–7 mm, villous abaxially and at margin; bracteoles shorter than the calyx. **Corolla**: tubular-campanulate, 22–30 mm, pale yellow, becoming dark blue when dry; villous at the base adaxially. Upper and lower lips with lobes, each 6–8 × 6–12 mm, entire and glabrous abaxially. **Carpel**: stigma sub-globose; style longer than filaments. **Stamens**: filament 14–16 mm, anthers long ovoid and densely villous with mucronate base. **Host**: reported on hosts from various families including Amaranthaceae, Tamaricaceae, Zygophyllaceae, Fabaceae and Rosaceae, e.g. *Ammopiptanthus mongolicus*, *Caragana tibetica*, *Potaninia mongolica*, *Reaumuria songarica*, *Salsola passerina*, *Tetraena mongolica* and *Zygophyllum xanthoxylum*. **Ecology and distribution**: north-central China to Inner Mongolia and Xinjiang, on grassy plains and slopes 1,000–2,400 m. **Key diagnostics**: among Chinese species, diagnostic traits are the 4-parted calyx, villous adaxial corolla base, and corolla drying dark blue. **Similar species**: *Cistanche lanzhouensis* is considered conspecific; it has a 5-parted calyx with lobes less equal, corolla drying pale brownish-yellow rather than blue.

Cistanche sinensis China
(left) dried specimen
(right) photo: Majed Aldughayman

NAME **Cistanche tubulosa** (Schenk) Wight ex Hook.f.
SYNONYMS **Cistanche calotropidis** (Edgew.) Beck; **Cistanche tinctoria** (Forssk.) Beck; **Phelypaea tubulosa** Schenk

Habit: robust, unbranched, erect, dense and many-flowered spike. **Height**: (15–)20–50(–130) cm. **Stem**: thick, glabrous, with numerous flowers crowded on a broad spike up to 9 cm in diameter. Stems yellow; lower scales sinuate, imbricate and broadly lanceolate, to 3 cm long; upper scales greyish, sinuate, ovate-lanceolate, 4–10 mm long. **Calyx**: tubular, 5-parted, about ½ the length of the corolla; lobes subequal, oblong, obtuse; floral bracts deeply sinuate, ovate-lanceolate, equal or slightly exceeding the calyx, grey, 14–22 mm long. Bracteoles sinuate, oblong-lanceolate, equalling or shorter than calyx, grey, 2–3 mm long with a membranous margin. **Corolla**: broadly tubular-campanulate, 34–52 mm long, with 5 equal lobes (scarcely 2-lipped), lemon yellow to deep yellow. **Carpel**: stigma bilobate, pale; style cylindrical, oblique, glabrous, protruding. **Stamens**: didynamous, epipetalous and filaments densely woolly at base; anthers cordate and densely woolly, with rounded base and acute apex. Rarely protrude from corolla. **Hosts**: various desert shrubs, especially in the Amaranthaceae and Tamaricaceae, including: *Haloxylon salicornicum*, *Capparis spinosa*, *Zygophyllum propinquum*, *Tamarix* spp. and *Salsola* spp. **Ecology and distribution**: desert flats, sand, gravel substrates, mudstone and seasonally arid saline habitats. Reportedly in East Africa and throughout the Middle East, east to India and north to Kazakhstan, but distribution not properly known. **Key diagnostics**: robust yellow spikes; desert ecology. **Similar species**: recent molecular data reveal that two yellow-flowered plants co-occur in the Middle East: entirely yellow forms are *C. tubulosa* (for which a neotype was assigned, collected in Sinai), and this plant is closely related to *C. violacea;* meanwhile plants with strong purple pigmentation, though morphologically similar to *C. tubulosa*, are genetically distinct; these are now referred to as ***Cistanche tinctoria*** (of which the type specimen was collected in Yemen).

Cistanche tinctoria
Israel (left)
Cistanche tubulosa
West Bank, Palestine (right, page 144)
Israel (page 145)

NAME ***Cistanche violacea*** (Desf.) Hoffmanns. & Link
SYNONYMS ***Cistanche hybrida*** Beck; ***Orobanche violacea*** (Desf.) Wallr.; ***Phelypaea violacea*** Desf.

Habit: medium-sized perennial, simple or rarely branched. **Height**: (9.5–)15–38(–50) cm. **Stem**: simple, fleshy and whitish, with a bulbous base, 25–50 mm thick. Inflorescence conical, dense, up to 35 cm and often beginning at ground-level; scales ovate-lanceolate, greyish white with scarious margin. **Calyx**: (10–)14–19.5 mm, glabrous, scarious surface, with 5 subequal oblong-ovate lobes. Floral bracts ovate-oblong, equalling or exceeding calyx length, (11–)15–20.3 × 4–8 mm. Bracteoles ovate-lanceolate, shorter than calyx at (7–)10–15 mm. **Corolla**: broadly tubular, slightly curved, erecto-patent, 15–18(–40) mm long with subequal lobes rounded at apex. Folds between anterior lobes descending to basal ⅓ of corolla. Whitish with deep violet staining or striation distally and with prominent yellow folds on interior surface. **Carpel**: stigma thick and sub-bilobed, whitish, sometimes with violet staining; style curved, glabrous, whitish but with some violet staining. **Stamens**: inserted into basal ⅓ of corolla tube, hairy at insertion point. Anthers ovate with a mucronate-acuminate tip and some hair at apex. **Hosts**: various desert shrubs from the Amaranthaceae, Plumbaginaceae, Zygophyllaceae and Tamaricaceae including, *Atriplex halimus*, *Hammada* spp., *Noaea mucronata*, *Anabasis articulata*, *Limoniastrum* spp. and *Tamarix* spp. **Ecology and distribution**: sandy desert and dry scrub habitats, with tolerance for saline coastal dunes and scrub. Reported across northern Africa to Mauritania and Chad, as well as the Levant and south to Saudi Arabia. **Key diagnostics**: glabrous spikes with white and purple colouration. **Similar species**: appears to hybridise with *C. tubulosa* s.l. and form intermediates (verified with molecular data); the lectotype is from Tunisia; all glabrous, pale yellow and violet entities have often been described as *C. lutea* (the type location of which is in nearby Algeria); it is unclear to what extent *C. violacea* and *C. lutea* are distinct. South Spanish populations have variably been described as *C. phelypaea* and *C. lutea*; further work is needed to resolve how the many names for plants belonging to this complicated group are applied.

Putative hybrids
Cistanche tubulosa × *C. violacea* Israel (page 150)
Cistanche aff. *lutea* Spain (page 151)

Cistanche violacea
Israel (right, page 148, 149)

Additional species (many doubtful): numerous dubious or poorly known species exist across the range of the genus.

NAME ***Cistanche aethiopica*** Beck

Habit: a cylindrical, lax spike, exceeding the scaly stem. **Height**: 40 cm. **Stem**: weak, swollen at the base, scaly, striated and glandular-pubescent; lower scales glabrous and ovate; upper scales longer, abundant, glandular-pubescent. **Calyx**: deeply segmented and anteriorly connate; bidentate, veined and pointed, glandular-pubescent; equalling the length of the corolla tube. **Corolla**: tubular, slightly enlarged towards the mouth; erecto-patent, straight along the back, teeth sub-parallel; 15–20 mm long; probably yellow (unclear from the holotype). Upper and lower lip lobes entire or emarginate, elongated, rounded. Lacinia all deeply unevenly denticulated, sub-glabrous, folded on the margin. **Carpel**: stigma bilobed, yellow; style short, glandular-hairy. **Stamens**: inserted 2–3 mm above the base of the corolla. Sparsely hairy, with glabrous anthers. **Host**: unknown. **Ecology and distribution**: apparently distributed across Ethiopia. Found roughly 960 m above sea level. Possibly in Eritrea. **Key diagnostics**: geography alone; a poorly known taxon. **Similar species**: unknown; a doubtful species.

NAME ***Cistanche afghanica*** Gilli

Habit: an erect, short, dense spike. **Height**: 5–8 cm. **Stem**: 3.5 cm across at the base, glabrous, incrementally scaly; scales narrowly triangular or linear-lanceolate, obtuse, the lowermost glabrous, and uppermost are downy-ciliate on the margins. **Calyx**: bell-shaped, glabrous tube, 5-lobed, lobes subequal, with a white-indumentum along the margin; 3 lobes semi-orbicular and rounded, 2 sub-triangular and acute, 10–13 mm long. Floral bracts narrowly triangular, 15–20 mm long, 3–5 mm wide. **Corolla**: tubular, strongly curved, apically expanded, externally glabrous, rather downward-pointed, 5-lobed; 25–30 mm; whitish beneath, distally violet-blue. **Carpel**: stigma suborbicular, subentire; style glabrous. **Stamens**: filaments glabrous; anthers mucronate, densely woolly. **Host**: unknown. **Ecology and distribution**: high elevation desert sands in Afghanistan (>1,300 m); more broadly, also reported from Soviet Central Asia. **Key diagnostics**: a doubtful taxon, likely to be part of the *C. salsa* complex. **Similar species**: as above, doubtfully a true species; reputedly differs from a further entity, *C. ambigua* (not covered here) in its calyx with 2 sub-triangular sharp teeth, calyx, bracts much longer and narrowly triangular, and bracteoles narrowly linear. A doubtful species.

NAME ***Cistanche compacta*** (Viv.) Bég. & A.Vacc.

A glabrous plant reported from the coast of north-east Libya that no doubt belongs to either of the wider yellow- or purple-flowered lineages in the region.

NAME ***Cistanche laxiflora*** Aitch. & Hemsl.

Habit: robust, long, many-flowered spikes, often with flowers to near ground-level. **Height**: to 1.1 m. **Stem**: dark grey; scales rather lax. **Calyx**: tubular, 19–26 mm long, ½ as long as or slightly longer than the corolla; calyx lobes 5, 3 longer, ovate, obtuse. Floral bracts ovate-lanceolate, obtuse, very sparsely hairy at the margins, exceeding the calyx, dark grey; floral bracteoles oblanceolate, acute, as long as calyx, sometimes shorter, grey-purple. **Corolla**: infundibuliform, tube white outside with 2 orange stripes inside, 40–50 mm; lobes 5, equal, with pale purple veins. **Carpel**: stigma white. **Stamens**: inserted near the corolla base, woolly. **Host**: *Tamarix* (Tamaricaceae); possibly other trees and shrubs. **Ecology and distribution**: deserts from the Middle East (recently reported in Israel) to China in its widest circumscription (but see below); Afghanistan is the type location. **Key diagnostics**: long, rather lax, pale spikes; stem scales, floral bracts and bracteoles oblong-lanceolate, entire; calyx with 3 obtuse lobes longer than the remaining 2; often on *Tamarix*. **Similar species**: records of *C. tubulosa* from China cultivated on *Tamarix* correspond best with *C. laxiflora* in its wider circumscription.

NAME ***Cistanche mongolica*** Beck

A whitish-violet entity parasitic on Tamaricaceae, e.g. *Tamarix ramosissima* and *T. hispida* reported from central Asia to western Mongolia in deserts correspond with *C. laxiflora*.

NAME ***Cistanche ridgewayana*** Aitch. & Hemsl.

A yellowish-white, purplish-brown-tinged plant with pubescent bracts and stamens and reduced bracteoles, reported on *Tamarix*, and on various Amaranthaceae and Fabaceae, in the deserts of Afghanistan, Iran, Turkmenistan and Uzbekistan. A doubtful taxon, probably close in form to *C. fissa*.

NAME ***Cistanche trivalvis*** (Trautv.) Korsh.

A glabrous plant parasitic on *Alhagi*, *Calligonum*, *Tamarix* and various Amaranthaceae including *Haloxylon* in desert scrub, in Turkmenistan, Tajikistan and Iran. The 3-valved seed capsule is the key diagnostic, but this feature occurs in other species so may not be stable. Reputedly close to (or synonymous with) *C. afghanica*.

NAME *Conopholis alpina* Liebm.

Habit: racemose, terminal inflorescences, colony-forming. **Height**: 8–28 cm. **Stem**: glabrous, simple, erect; rather thick, yellowish-cream; scales 1.2–2.1 cm long, 0.3–0.7 cm across, glabrous, lanceolate to narrowly elongate triangular. **Calyx**: cylindrical, 0.33–0.45 cm, irregularly 2- or 4- to 5-toothed or lobed with acute to obtuse teeth; floral bracts glabrous, lanceolate to narrowly elongate triangular, often concealing the calyx. **Corolla**: narrowly tubular-campanulate; 0.75–1.55 cm long; yellowish-white with brown markings. **Carpel**: stigma exserted, deciduous at maturity; style 5.5–8 mm long (incl. stigma). **Host**: parasitic on various species of *Quercus* (Fagaceae; predominantly *Q. alba*). **Ecology and distribution**: distributed from the south-east USA to Panama in deciduous woods and mixed montane forests, at 1,500–2,800 m. Blooms February– August depending on distribution and altitude. **Key diagnostics**: colonies of up to 25 inflorescences on the roots of oak in deciduous woods. **Similar species**: similar to *C. americana*, which occurs at lower elevation <600 m, is more orange and has longer corollas and persistent styles.

Infraspecific taxa

Conopholis alpina* var. *mexicana is distinguished by the lack of veins on the bracts, which scarcely conceal the calyx; pubescence mostly glandular.

Conopholis alpina New Mexico, USA (above, right) photos: Donald R. Tomberlin

NAME ***Conopholis americana*** (L.) Wallr.
SYNONYM ***Orobanche americana*** L.

Habit: erect, colony-forming. **Height**: 6–20 cm. **Stem**: 0.4–1.2 cm across, creamy yellow to tan-brown, glabrous, simple or sparingly branched, erect; scales 0.55–1.8 cm long, 0.45–1.2 cm wide, glabrous, with sessile glands, or rarely minutely glandular-pubescent along the margins, broadly to narrowly lanceolate or narrowly triangular, widest at or just above the base, apex acute to nearly rounded. **Calyx**: irregularly 4- to 5-toothed or lobed, the tube ± cylindrical, split anteriorly from middle to base, the teeth or lobes acute; bracteoles 2. Calyx 35 mm long. Floral bracts 0.55–1.8 cm long, 0.2–0.8 cm wide, glabrous or rarely minutely pubescent along margins, broadly to narrowly lanceolate or elongate triangular, often nearly or wholly concealing the calyx. **Corolla**: tubular; 0.8–1.4 cm long. Upper lip shallowly curved downwards over the lower lip. Lower lip 3-parted; creamy-white. **Carpel**: stigma rather thick, pale; style: 0.5–1.3 cm long (incl. stigma), persistent, reflexed downwards. **Stamens**: protruding above the upper lobe. **Host**: parasitic on *Quercus* and *Fagus* (Fagaceae). **Ecology and distribution**: restricted to eastern North America from Nova Scotia to Michigan and Wisconsin and to Florida under oaks in moist, deciduous or mixed woods, at 30–600 m. Flowers from mid-February in the south to mid-June in the north. **Key diagnostics**: 'cob of corn'-like appearance, colony-forming under oak. **Similar species**: similar to *C. alpina* but with a shorter corolla tube and persistent, reflexed style and less mountainous habitats.

Conopholis americana (in fruit) Pennsylvania, USA (above, right) photos: Tony DeSantis

NAME ***Epifagus virginiana*** (L.) Barton
SYNONYM ***Orobanche virginiana*** L.

Habit: very slender, branched, lax spikes. **Height**: (3.5–)15–45(–54) cm. **Stem**: racemes held on 1–12 branches, slender; stems yellow, brown or purplish, swollen at base into corm-like structures; scales triangular, ovate or linear, 2–5 mm long. Pedicels 1–3 mm and bracteoles paired, adnate to base of calyx, triangular-ovate and 1–2 mm long. **Calyx**: 2–3 mm long, with acuminate lobes to 1 mm. **Corolla**: narrowly tubular, laterally compressed, 8–12 mm long, cream with purple longitudinal bands. Cleistogamous flowers often present on lower ½ of the raceme, with corolla closed at apex and only around 3 mm long. Upper lip erect with entire or slightly notched margin (or incurved slightly). Lower lip lobes erect and somewhat spreading with acute apices. **Carpel**: stigma pale yellow, usually slightly exserted from corolla; style glabrous. **Stamens**: included or barely exserted from corolla, inserted in lowest ⅓ of corolla interior; anthers sagittate with acuminate tip. **Host**: *Fagus grandifolia* (Fagaceae). **Ecology and distribution**: deciduous forests in the eastern half of North America, north to Quebec and Ontario, south to northern Mexico. Flowering July–October. **Key diagnostics**: slender, highly branched stems. **Similar species**: none.

Epifagus virginiana Pennsylvania, USA (above, right) photos: Tony DeSantis, Connecticut, USA, (page 160, 161) photos: Eric Powell

NAME ***Eremitilla mexicana*** Yatsk. & J.L.Contr.

Habit: low, dense, spirally imbricate, ovoid clusters of flowers at ground-level (stem above-ground absent). **Height**: 15–45 cm. **Stem**: unbranched, succulent, 6–25 mm across, cylindrical, with dense, squamiform, overlapping, broadly oblong to depressed-ovate scales 4–6 × 5–8 mm. **Calyx**: with 4–5 sepals, 18–25 × 9–15 mm, adaxially glabrous, abaxially densely vesicular-hairy. Floral bracts 21–28 × 14–18 mm, ovate to elliptic above a short, broad, stalk-like base, reddish-brown. Bracteoles usually 2, aligned with lateral sepals, 14–21 × 7–12 mm, narrowly elliptic to narrowly rhombic. **Corolla**: 4-lobed (by almost complete fusion of the 2 upper lobes of a 5-merous corolla), glabrous, cream to pale yellow, abruptly constricted and reddish-brown distally, densely pubescent externally with orange hairs; tube 24–28 mm; lobes 10–13 mm; upper lobe twice as wide as the others. **Carpel**: stigma discoid-crateriform, 2.0–2.5 mm diameter, dull yellow; style 5–8 mm, not exserted. **Stamens**: not exserted, filaments free, inserted at the ovary base, 14–18 mm, glabrous. **Host**: *Hedyosmum mexicanum* (Chloranthaceae). **Ecology and distribution**: endemic to the inland slopes of the Sierra Madre del Sur in Guerrero in Mexico, at elevations of 1,800–2,000 m, in the seasonally dry forests including deciduous woods and hardwood montane forests. Flowering in June and July. **Key diagnostics**: 5-ribbed ovaries and unusual stamens with anthers partially embedded in an expanded filament. **Similar species**: none.

Eremitilla mexicana
illustrated (right)

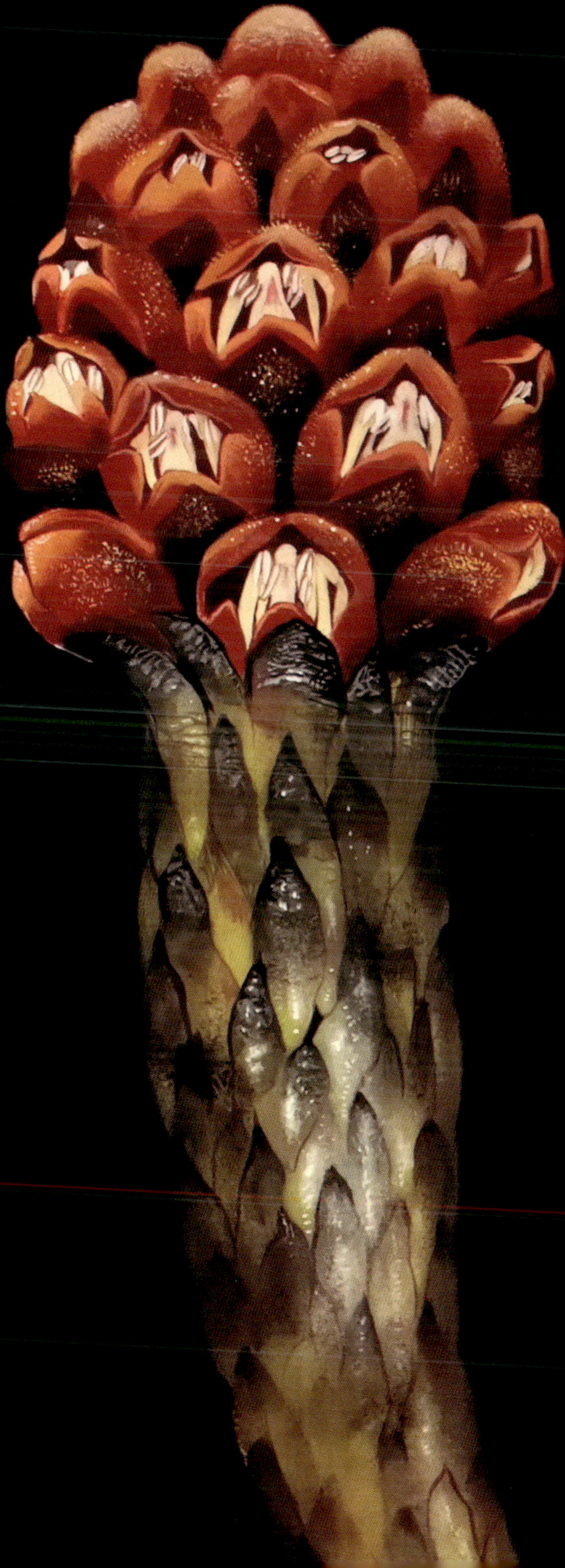

NAME ***Gleadovia konyakianorum*** Odyuo, D.K.Roy & Aver.

Habit: short clusters erupting immediately around host stem base; flowers borne singly on each stem apex. **Height**: to 10 cm. **Stem**: fleshy, yellowish with spirally arranged scales; scales lanceolate, oblanceolate to ovate, smooth, glabrous. **Calyx**: clasping, glabrous, split to nearly ⅓ of its length with 5 irregularly elliptic-lanceolate lobes apically, 4.0–5.5 × 1–1.5 cm (shorter than corolla tube). **Corolla**: tubular, strongly constricted at the mouth so as to appear partially closed; somewhat inflated below; 5.5–7.5 × 1.0–1.2 cm, white with translucent veins. Upper lip with 2 lanceolate lobes, apically acute. Lower lip with 3 lanceolate lobes, apically acute. **Carpel**: stigma narrowly subulate, 2.3–2.5 cm long, incurved, 2-lobed, lobes parallel, glabrous; style stout, glabrous. **Stamens**: narrowly subulate, incurved with 2 parallel lobes, glabrous. **Host**: shrubby *Strobilanthes* spp. (Acanthaceae). **Ecology and distribution**: found in Mon District, Nagaland, India, on the banks of streams in semi-evergreen mountain forests, at 1,500–1,600 m. Flowering in April. **Key diagnostics**: white flowers; urceolate calyx and flowers borne singly. **Similar species**: similar to others in the genus but with strictly 1-flowered inflorescences (appearing clustered).

Gleadovia konyakianorum
illustrated (right)

NAME ***Gleadovia ruborum*** Gamble & Prain

Habit: few flowers borne at ground-level in dense corymbose or paniculate inflorescences. **Height**: 10–18 cm high (including underground stem). **Stem**: above-ground stems absent; below-ground thick, yellow with imbricate bracts ca. 1.5 cm long, sheathing, blunt, shiny. **Calyx**: yellow, tinged pink, tubular, somewhat inflated with equally 5-lobed, lobes rounded, divided to less than ½ the tube length, 2.5–3 cm long. **Corolla**: tube much longer than the calyx, slightly curved, 2-lipped, to 5 cm long; white at base, deep reddish-pink towards the apex with darker longitudinal veins. Upper lip connate, with rounded lobes. Lower lip with 3 narrow, acute lobes. **Carpel**: stigma with 2 fleshy, semi-orbicular lobes depressed in the centre; style shorter than the filaments. **Stamens**: with filaments bent at point of insertion; anthers spurred, connectives produced beyond the anther lobes, 3-fid at the apex. **Host**: *Rubus niveus* (Rosaceae). **Ecology and distribution**: Himalayas (northern Pakistan/India and Tibet) and southern China (south-west Yunnan to west Hunan), on northern slopes in shaded forests dominated by *Cedrus deodara*, *Abies pindrow* and *Quercus floribunda*, at 2,500–3,000 m. Flowers June–July; in southern China, found in temperate rainforest under bamboo in humid forests and thickets, 900–3,500 m. **Key diagnostics**: rose-pink flowers and parasitic on *Rubus*. **Similar species**: easily identified based on the characters stressed above.

Additional species

NAME ***Gleadovia banerjiana*** Deb

Recognised by its flesh, cymose inflorescence of 1–3, white to rose-tinted flowers 5–5.5 × 3–4.5 cm with incurved capitate, 2-lobed, stigmas. Parasitic on *Strobilanthes discolor* in Assam, India.

NAME ***Gleadovia mupinensis*** Hu

A poorly known species with few flowers borne in corymbs, ± spaced on the upper part of stem; bracteoles linear-lanceolate; pedicel 2–9 cm long; calyx tubular, slightly enlarged upward; flowers lavender to pale purple, pale purple-red (rarely white); corolla lobes densely villous on both surfaces. Found in Sichuan, central China in forests, on roadsides, and other moist places, at altitudes of 3,000–3,500 m. Similar to *G. ruborum* but a darker purple corolla.

Gleadovia ruborum
illustrated (right)

NAME ***Harveya bodkinii*** Hiern

Habit: erect, congested, conspicuous red inflorescences borne at ground-level with 4–10(–18) rather large flowers; glandular-pubescent throughout, especially at the apex. **Height**: up to 15(–23) cm. **Stem**: stem above-ground absent (subterranean stem not observed); scales orbicular, to 5 mm. **Calyx**: tubular-campanulate, broadly cylindrical, concolourous with corolla tube. Floral bracts 10–14 mm long, lanceolate to ovate; bracteoles 2, opposite, 10–14 mm, linear-lanceolate (often inserted in the calyx). **Corolla**: tube orange to red, darker distally, narrow, erect; limb scarlet or orange-red, yellow in the throat, bilabiate, 47–52 mm long, flat to slightly oblique with lower lip jutting slightly forward, 2 upper lobes subequal or slightly smaller than the 3 anterior lobes, lobes orbicular, 1.0–1.5 cm across. **Carpel**: stigma exserted, yellow; style recurved. **Stamens**: inserted halfway from the corolla base. **Host**: unknown. **Ecology and distribution**: high elevations of the Cape provinces of South Africa. **Key diagnostics**: clumped, congested inflorescences with large, scarlet flowers borne at ground-level; mountain ecology. **Similar species**: similar to *H. bolusii* but less elongated and differing in the characters stressed above.

Harveya bodkinii South Africa (above, right) photos: Mark Johnston

NAME ***Harveya bolusii*** Kuntze

Habit: lax, erect, slender terminal spikes of rather few flowers from near ground-level. **Height**: up to 25 cm. **Stem**: maroon to deep red, simple or rarely branching at the base, glabrous-pubescent below to pubescent or villose above. **Calyx**: tubular-campanulate, broadly cylindrical, longer dorsally; 1.2–2.5 × 0.5–0.7 cm. Floral bracts pubescent to villose, 0.7–1.3(–1.9) × 0.3–0.5 cm, elliptic-lanceolate, obtuse or subacute at apex, scale-like, and subsessile-sessile, often larger than the stem bracts. **Corolla**: narrowly funnel-shaped to cylindrical, rather strongly curving upward; limb bilabiate; lobes orbicular, 0.5–0.9 cm across, upper and lower lobes subequal or the lower lobes slightly larger, margins minutely crenate or entire, ciliate; (2.4–)2.7–3.5(–3.9) cm long, tube 0.2–0.3 cm across at the base. Tube red to orange with yellow throat; bright yellow on ventral side of the tube and inside throat. **Carpel**: stigma linear-oblong, not capitate, 0.3–0.4 × 0.1–0.2 cm; style glabrous, usually exserted, strongly incurving below stigma. **Stamens**: inserted (0.9–)1.1–1.5(–1.7) cm from base of corolla, nearly exserted; filaments flattened and glabrous; anthers glabrous. **Host**: various Asteraceae including *Stoebe* spp., *Disparago* spp., *Euryops abrotanifolius*, as well as *Cliffortia* spp. (Rosaceae), and reportedly, various Cyperaceae. **Ecology and distribution**: Cape provinces of South Africa at the crests of mountain ranges on south- or west-facing mountain slopes. Flowers November–January. Bird-pollinated. **Key diagnostics**: lax, bright orange spikes of arched, long-tubular flowers. **Similar species**: superficially similar to *H. scarlatina*, which has shorter spikes, and different ecology and host range.

Harveya bolusii
South Africa (right)
photo: Mayur Prag

NAME ***Harveya capensis*** (Burm.f.) Hook.
SYNONYMS ***Harveya lactea*** C.Presl; ***Orobanche capensis*** Burm.f.

Habit: tall, erect, lax spikes of large, conspicuous flowers. **Height**: 20–31(–50) cm. **Stem**: dull brown, terete or obtusely angled in cross-section, erect, long and flexuous, fleshy or woody below, simple or branching most often below-ground. **Calyx**: tubular-campanulate, 1.7–2.2 × 1.0–1.3 cm. Floral bracts scale-like, often larger than caudal leaves, but similar in shape, position, colour. **Corolla**: narrowly funnel-shaped with large petals; tube laterally compressed. Limb bilabiate, oblique, 1.9–3.7 cm in diameter. Upper lip lobes smaller, ± erect; lower lip lobes spreading; all with or without wavy margins; tube white to cream-coloured or pale yellow; lobe margins sometimes pinkish, blackish when bruised or with age. **Carpel**: stigma entire or slightly lobed, capitate or not; greenish; style exserted, incurved. **Stamens**: inserted or exserted, filaments terete or flattened, glabrous; filaments inserted 1.3–1.6 cm from the base of corolla. **Host**: *Erica mucosa* (Ericaceae), *Hydrocotyle* spp. (Araliaceae), *Centella* spp. (Apiaceae), *Anthospermum* spp. (Rubiaceae), *Cliffortia* (Rosaceae), *Searsia* (Anacardiaceae). **Ecology and distribution**: Cape provinces of South Africa near seasonally wet mountain slopes on rocky sandstone substrates, near seepages and mountain springs. Flowers November to January. Moth-pollinated. **Key diagnostics**: large, white, near-actinomorphic flowers with wavy-edged lobes. **Similar species**: similar to *H. purpurea* but with corolla tube laterally strongly compressed.

Harveya capensis
South Africa (left, right)
photos: Mayur Prag

NAME ***Harveya comorensis*** Vatke

Habit: a small, low, pale plant with few flowers; short-pubescent throughout. **Height**: to 12 cm. **Stem**: subterranean stem pale and slender with numerous bracts; stem above-ground virtually absent. **Calyx**: cream with darker netter veins, fused, with short triangular teeth. **Corolla**: flowers sub-erect; tube constricted, bent at the throat, narrowly campanulate, 2.5 cm long, 2.7 cm across, pale whitish-yellow, with pale yellow-orange in the throat; lobes orbicular, white, recurved (especially the upper lip lobes). **Carpel**: unobserved. **Stamens**: unobserved. **Host**: unknown. **Ecology and distribution**: known only from the forests of Comoros and Madagascar at elevations of ca. 1,000 m. **Key diagnostics**: a poorly known taxon, distinguished by its small, low habit, and few, white, yellow-throated flowers. **Similar species**: none.

Harveya comorensis
Mayotte, Comoro Islands (left, right)
photos: Ehoarn Bidault
(Missouri Botanical Garden)

NAME ***Harveya huttonii*** Hiern
SYNONYMS ***Harveya leucopharynx*** Hilliard & B.L.Burtt; ***Harveya pulchra*** Hilliard & B.L.Burtt; ***Harveya silvatica*** Hilliard & B.L.Burtt

Habit: tall, erect, often clumped, lax spikes of large, conspicuous flowers along ½ the stem axis. **Height**: 15–40(–50) cm. **Stem**: greenish-yellow, or dull red to crimson, simple or scarcely branched, below-ground simple to heavily branching with up to 20 flowering stems arising from single rhizome. **Calyx**: broadly tubular, mouth somewhat oblique, base rounded, brownish-red, lobes 5; 1.4–2.4(–3.0) cm × 0.7–1.1(–1.5) cm. Floral bracts opposite or rarely alternate, 1.4–2.8 × 0.3–0.7 cm, elliptical or oblanceolate, concave, pubescent to villose, acute. **Corolla**: narrowly funnel-shaped, the lower ⅓ of the tube abruptly contracted and slightly curved, 2.8–4.5 cm long. Upper lip lobes suborbicular, 0.8–1.2 × 1.2–1.6 cm, lower lip lobes 1–2 × 1.2–2 cm, narrow at the base; all lobes wavy to crenulate, or lacerate, and minutely to heavily ciliate at the margins. Corolla throat pale or bright yellow-white; limb pale or bright pink or lilac. **Carpel**: stigma capitate and grooved, partially occluding the throat, just inserted; style 2.5–3.3 cm long, abruptly bent inwards near tip. **Stamens**: included; filaments inserted 5–13 mm from the base of corolla, glabrous-puberulent at point of insertion; anthers glabrous. **Host**: various, including *Anthospermum* spp. (Rubiaceae), *Felicia* spp. (Asteraceae), *Euryops tysonii* (Asteraceae) and *Sparrmannia ricinocarpa* (Malvaceae). **Ecology and distribution**: subtropical regions of South Africa, Swaziland and Lesotho, at 1,900–3,000 m, on damp, south-facing, grassy slopes and stream gullies, montane grassvelds and forest margins. Flowers November–February. Insect-pollinated. **Key diagnostics**: large, pink, yellow-throated, near-actinomorphic flowers with wavy-edged lobes. **Similar species**: similar to *H. purpurea* and co-occurring on the Eastern Cape but with different calyx and throat shape; throat often yellow in *H. huttonii* (rather than yellow spot on the palate of the lower lip only in *H. purpurea*).

Harveya huttonii South Africa (above, right) photos: Mayur Prag

NAME *Harveya hyobanchoides* Schltr. ex Hiern

Habit: broad, erect, sub-conical, congested spikes of 12–36 sessile flowers. **Height**: 18–34 cm. **Stem**: fleshy, villose, simple, or rarely branching below, 0.8–1.3 cm in diameter; stem with viscid hairs; scales red. **Calyx**: campanulate at the base, cylindrical above, with shallowly divided calyx lobes; (2.1–)2.3–3.0 × (0.5–)0.7–1.1 cm. Floral bracts concave, adpressed, (1.5–)1.8–3.0(–3.3) × 0.7–1.0 cm, elliptical to oblanceolate, villose to densely villose abaxially, pubescent adaxially, acute, sessile. **Corolla**: tubular with short, reflexed lobes; tube cylindrical, 0.4–0.6 cm diameter at base, gradually inflating to 0.7–1.1 cm at throat, longer dorsally than ventrally; limb 1.0–1.4(–1.7) × 1.0–1.4 cm, bilabiate, all lobes slightly reflexed, approximately equal, 0.4–0.6(–0.9) × 0.3–0.5 cm, semiorbicular to deltoid, villose in and out, somewhat concave, entire. Corolla tube yellow, distally grey-green. **Carpel**: stigma sub-globose, yellowish. **Stamens**: arising from a ring of dense hairs in corolla throat, filaments flattened. **Host**: *Aspalathus uniflora* (Fabaceae). **Ecology and distribution**: South Africa, Lesotho and Swaziland. Found on sandy soils and coastal substrates. Flowers July to October. Insect-pollinated. **Key diagnostics**: calyx divided nearly to the base, with 5 very unequal lobes. **Similar species**: superficially very similar to *H. roseoalba,* differing subtly in corolla colour and in host range.

Harveya hyobanchoides South Africa (above, right) photos: Abdul-lateef Ismail

NAME ***Harveya pauciflora*** (Benth.) Hiern
SYNONYMS ***Aulaya coccinea*** Harv.; ***Aulaya pauciflora*** Benth.; ***Harveya coccinea*** (Harv.) R.Schlechter; ***Harveya tubulosa*** Harv. ex Hiern

Habit: slender, erect or flexuous-ascending herb with 4–6 flowers borne in long, lax racemes. **Height**: up to 48(–60) cm. **Stem**: dull red, markedly glandular-pubescent above, angular-furrowed, dull red to dusky purple stem and scales. **Calyx**: campanulate-oblong, deeply divided; lobes lanceolate, obtuse or scarcely acute, 1.0–1.2 cm long, purplish-red. **Corolla**: broadly tubular, very densely pubescent, with short, rounded lobes; 1.9–2.1 cm long; tube cream to pink, limb pink. **Carpel**: stigma slender, oblong-clavate. **Stamens**: glabrous. **Host**: *Ficinia* spp. (Cyperaceae). **Ecology and distribution**: South Africa, at 300–600 m; distribution improperly known. **Key diagnostics**: all aerial parts densely white-hairy. **Similar species**: none.

Harveya pauciflora
South Africa (left, right)
photos: courtesy of Groeneweide Farm

NAME ***Harveya pumila*** Schltr.
SYNONYMS ***Harveya crispula*** Conrath; ***Harveya randii*** Hiern

Habit: spicate or corymbose clusters of flowers at ground-level (appearing stemless). **Height**: up to 12 cm. **Stem**: virtually absent above-ground; stems fleshy, 0.3–0.6 cm in diameter, rarely branching, but if so, at or above the first flowering node, sub-glabrous. **Calyx**: campanulate at the base, campanulate to cylindrical above, with subacute, unequal lobes. Floral bracts concave, adpressed or antrorse, 0.9–1.4 × 0.4–0.6 cm, obovate to ovate, sessile, acute to obtuse at the apex, villose or grading from puberulent below to villose at the apex. **Corolla**: tube not compressed laterally; lobes ± equal, (0.5–)0.6–1.0 × 0.7–1.3 cm, orbicular to obovate, crenulate or rarely lobulate, often ciliate at margins; tube 3.5 cm long, externally cream to pink; lobes pale to deep pink; throat whitish with a small, yolk-yellow spot. **Carpel**: stigma white. **Stamens**: inserted at point of corolla inflation, filaments densely villose below. **Host**: *Polycnemum* spp. (Acanthaceae), *Metalasia* spp. (Asteraceae), *Anthospermum rigidum*, *A. aethiopicum* and *A. pumilum* (Rubiaceae). **Ecology and distribution**: Mozambique, south to South Africa in grassveld or rocky slopes and dry interior plateau regions. Flowers August–December. Insect-pollinated. **Key diagnostics**: easily distinguished by the ground-dwelling habit and the transverse band of dense hairs encircling the interior of the corolla at the point of anther insertion. **Similar species**: none.

Harveya pumila South Africa (above) photo: Andrew Hankey, (right) photo: Athol Ferguson

NAME ***Harveya purpurea*** Harv. ex Hook.

Habit: tall, erect, lax spikes of large, conspicuous flowers, usually over more than ½ the stem axis. **Height**: (8–)13–16(–34) cm. **Stem**: red, sometimes tinged green or yellow, rare branched, glabrescent to pubescent below, villose above, viscid, 0.2–0.4 cm in diameter, terete to obtusely angled. **Calyx**: strongly 10-nerved, campanulate to obconical at the base, broadly campanulate above, strongly 2-lipped, (1.6–)1.8–2.4(–2.7) × 0.7–1.3(–1.4) cm. Floral bracts opposite, elliptical-lanceolate, 1.0–1.3 × 0.2–0.4 cm, acute or rarely subacute, pubescent to villose. **Corolla**: tube inflating abruptly 3–5 mm from the base, gently curved, dorsally keeled, (2.3–)2.4–3.6(–3.8) cm in length, 0.9–1.5(–2.2) cm in diameter; tube white to yellow, limb pink to purple with a sulphur yellow spot on the palate. **Carpel**: stigma partially occluding the mouth of the corolla, capitate, usually globose (0.2–0.3 cm in diameter), occasionally oblong (0.6 × 0.3 cm). **Stamens**: inserted in the corolla tube at point of inflation (0.3–0.8 cm from the base); filaments 2.0–2.4 cm in length, puberulent-pubescent; anthers glabrous. **Host**: reported on various genera including *Roella* spp., *Wahlenbergia* spp., *Prismatocarpus diffusus* (Campanulaceae), *Myrsine* spp. (Primulaceae), *Erepsia ramosa* (Aizoaceae), *Chrysopogon* spp. (Poaceae), *Anthospermum* spp. (Rubiaceae) and *Stoebe* spp. (Asteraceae). **Ecology and distribution**: Cape provinces of South Africa in fynbos, rhenosterveld habitats and coastal plains. Insect-pollinated. **Key diagnostics**: large, rose-pink, yellow-throated, near-actinomorphic flowers with wavy-edged lobes. **Similar species**: similar to *H. huttonii* and co-occurring on the Eastern Cape but with different calyx and throat shape, and with yellow spot on the palate of the lower lip only.

Harveya purpurea South Africa (above left, above right, right) photos: Mayur Prag

NAME ***Harveya roseoalba*** J.C.Manning & Goldblatt

Habit: broad, erect, sub-conical, congested spikes of 25–50 sessile flowers. **Height**: up to 20 cm. **Stem**: robust, densely red woolly-viscid; scales deep pink-red, 15–25 mm across, imbricate below. **Calyx**: campanulate, deeply 5-lobed, all lobes narrow, with one lobe much-exceeding the other four; 25–30 mm. Floral bracts sessile, adpressed, obovate-rhomboidal, concave, 15–25 × 10–15 mm, fleshy, densely villous abaxially, glabrous adaxially; bracteoles arising from base of calyx or on pedicel if present, opposite, linear-oblanceolate, 20–25 × 3–4 mm, densely villous abaxially, glabrous adaxially. **Corolla**: narrowly funnel-shaped with short reflexed petals; tube concealed by bracts, pale cream (ageing pink), somewhat darker yellow along the distal exterior tube; upper lip 2-lobed and lower lip 3-lobed, all lobes subequal, patent, suborbicular, 5–6 × 5–6 mm, glandular-hairy on both sides with shorter hairs on adaxial surface; tube to 40 mm. **Carpel**: stigma oblong, pale; style strongly recurved. **Host**: *Hermannia velutina* (Malvaceae) and *Phylica laevigata* (Rhamnaceae). **Ecology and distribution**: rare and local in the Cape provinces of South Africa; rocky, semi-desert slopes, montane sandstone and clay-based fynbos. **Key diagnostics**: squat, sub-conical, spiciform, dark pink inflorescences with contrasting pale cream flowers partially obscured by bracts. **Similar species**: superficially like *Hyobanche* (and shares a 1-lobed anterior calyx lip) but with very different corolla morphology; similar to *Harveya hyobanchoides*, differing mainly in colour and host range.

Harveya roseoalba
South Africa (left, above, page 188, 189)
photos: Nicola van Berkel
illustrated (right)

NAME **_Harveya scarlatina_** (Benth.) Hook. ex B.D.Jacks.

Habit: lax, erect, slender terminal spikes of rather few flowers from near ground-level. **Height**: stem 11–20 cm. **Stem**: covered in red or crimson scales, often branching below, yellowish to greenish-yellow, viscid, puberulent, pubescent, or villose towards the apex, 0.5–0.6 cm in diameter. **Calyx**: obconical or cuneate below, broadly or narrowly cylindrical above, exceeding ½ the length of the corolla tube; (3.0–)3.2–4.2(–5.2) × 0.9–1.7 cm. **Corolla**: tube cylindrical, erect, scarcely inflated at the base; upper and lower lip lobes broadly triangular-orbicular to ovate, 0.8–1.2(–1.5) × 0.8–1.3 cm, the upper lobes strongly reflexed; tube limb pale yellowish-orange to orange-red, interior lobes dark orange; throat yellowish. **Carpel**: stigma clavate, not capitate, 0.3–0.4 × 0.1–0.2 cm; style somewhat or strongly exserted and strongly incurved at the apex. **Stamens**: somewhat exserted, filaments puberulent; anthers glabrous, fertile anthers crescent-shaped (0.3 × 0.2 cm), sterile anthers subulate (0.3–0.4 × 0.1 cm). **Host**: Asteraceae (*Euryops* spp. and *Oedera acerosa*). **Ecology and distribution**: Lesotho and regions of South Africa in high alpine grassveld among stunted shrubs and boulders. Flowers November–March. Bird-pollinated. **Key diagnostics**: erect, long-tubular, bright orange flowers borne in rather short terminal spikes. **Similar species**: superficially similar to *H. bolusii*, which has longer, more extended spikes, and different ecology and host range.

Harveya scarlatina
South Africa (left, right)
photos: Mayur Prag

NAME ***Harveya speciosa*** Bernh.
SYNONYMS ***Cycnium tubatum*** Benth.; ***Harveya anisodonta*** C.A.Sm.; ***Harveya cathcartensis*** Kuntze

Habit: tall, erect, lax spikes of rather few, large, conspicuous flowers. **Height**: (18–)30–50(>100) cm tall. **Stem**: erect, slender, unbranched, glabrous, 0.4–2.0 cm in diameter, brownish. **Calyx**: campanulate-obconical at the base, cylindrical above, strongly 10-nerved; lobes deltoid, acute, with shallow sinuses. Floral bracts opposite, elliptic-ovate or rarely spathulate. **Corolla**: tube very long (up 9 cm) and narrow, strongly erect then sharpy curved above; upper and lower lobes similar, all broadly orbicular, spreading up to 0.7 cm in diameter, somewhat reflexed and weakly crenate; tube dull yellowish externally; lobes white, the throat tinged yellow; corolla with indumentum. **Carpel**: stigma pale yellow, glabrous, capitate to spathulate; style glabrous, incurved above, extending into the corolla throat. **Stamens**: short, 0.3–1.0(–1.5) cm, equal or subequal, inserted below point of corolla inflation; anthers glabrous and essentially monothecous. **Host**: *Berkheya purpurea* (Asteraceae). **Ecology and distribution**: South Africa, in moist grasslands on mountain slopes and coastal plains. Flowers October–March. Moth-pollinated. **Key diagnostics**: large white flowers held erect on tall, slender stems; glabrous except for the corolla. **Similar species**: similar to *H. capensis*, which is hirsute throughout; *H. speciosa* has highly reduced anthers.

Harveya speciosa South Africa (above, right) photos: Mayur Prag

NAME ***Harveya squamosa*** Steud.
SYNONYMS ***Orobanche squamata*** G.Don; ***Orobanche squamosa*** Thunb.

Habit: robust, congested, pyramidal spikes with flowers from ground-level; often densely clump forming. **Height**: (20–)30–40 cm. **Stem**: erect, simple or with up to eight branches from the base, orange or red-orange. **Calyx**: campanulate at base, cylindrical above, not strongly bilabiate. Lobes lanceolate, sinuses between lobes varying. Floral bracts adpressed, lanceolate-obovate, acute at apex. **Corolla**: tube erect, narrowly tubular-funnel-shaped; limb scarcely 2-lobed; lobes imbricate, concave, orbicular to semi-orbicular. Corolla tube (3.2–)3.5–4.5(–4.8) cm long, 0.3–0.5 cm in diameter at base; tube limb and lobes concolourous orange (or external tube somewhat paler or yellowish); all with parts conspicuous short, yellow indumentum. **Carpel**: stigma oblong-clavate, 0.5–0.6 × 0.1–0.3 cm, scarcely capitate, yellow. **Stamens**: inserted approximately halfway from the base of the corolla tube; filaments not flattened; anthers glabrous. **Host**: various hosts reported included: *Arctotis decurrens* and *Othonna leptodactyla* (Asteraceae), *Aspalathus* spp. (Fabaceae), *Cliffortia* spp. (Rosaceae), *Ehrharta* spp. (Poaceae), *Scirpus* spp. (Cyperaceae), and *Willdenowia* spp. (Restionaceae). **Ecology and distribution**: south-western South African Cape provinces on coastal dunes. Flowers September–November. Bird-pollinated. **Key diagnostics**: bright orange colouration, habit and coastal ecology. **Similar species**: somewhat similar to *H. bolusii*, but *H. squamosa* has generally more uniform yellow to orange-red corolla, denser, many-flowered spikes and distinct ecology.

Harveya squamosa South Africa (above, page 197) photos: Sebastian Hatt, (page 196) photo: Mayur Prag, illustrated (right)

NAME ***Harveya stenosiphon*** Hiern

Habit: tall, slender herb with few, large flowers, scattered along the stem. **Height**: up to 60 cm. **Stem**: unbranched or rarely branching, gradually becoming villose at apex; stem, scales, bracts and calyces maroon or greenish with maroon markings. **Calyx**: ± concolourous with stem; markedly pubescent. **Corolla**: tube narrowly cylindrical, inflating abruptly below; limb broad, slightly oblique, 2.9–5.0(–6.0) cm in diameter; lobes flat; corolla yellow to orange or scarlet. **Carpel**: stigma sub-globose, yellow, exserted. **Stamens**: inserted at point of corolla inflation; filaments flattened and glabrous; anthers glabrous. **Host**: *Osmitopsis osmitoides* (Asteraceae). **Ecology and distribution**: Cape provinces of South Africa in wet habitats with dense vegetation on south-facing ravines. Flowers November–March. Bird-pollinated. **Key diagnostics**: large, upward-facing brightly coloured flowers borne on long, slender stems. **Similar species**: none.

Harveya stenosiphon South Africa (above, right) photos: Mayur Prag

Additional species

NAME ***Harveya alba*** Hepper
SYNONYM ***Alectra alba*** (Hepper) B.L.Burtt
Habit: very lax, few-flowered spikes. **Height**: 10–40 cm. **Stem**: simple, erect above, decumbent below, glabrous; scales lanceolate, 0.2–0.6 × 0.2 cm, acute. Flowers opposite, in a lax terminal raceme. **Calyx**: campanulate with reflexed teeth; tube glabrous (sometimes with few hairs on the main veins); 0.6–0.9 × 0.4–0.6 cm, with teeth 0.3 cm. **Corolla**: tube erect, narrowly campanulate, 1.2 × 0.5 cm, with unequal lobes 0.3–0.8 cm long; margins entire; tube 1.2 × 0.5 cm across, 1.2–1.3 cm long; tube exterior pale yellowish-white, otherwise flowers complete white. **Carpel**: stigma linear with lateral stigmatic surfaces along the sides and one of the unequal anther thecae with very short point. **Stamens**: ± equal in length, anther sheaths unequally pointed and very short; filaments inserted ⅓ up corolla tube, filaments ± 0.6 cm long; anthers 2-thecate, all similar, rounded. **Host**: unknown. **Ecology and distribution**: wet, shaded montane forests, on the forest floor in leaf litter, at 1,500–2,200 m, in Burundi, Kenya, Rwanda and Uganda. **Key diagnostics**: lax, white, few-flowered, somewhat decumbent spikes found in leaf litter. **Similar species**: similar to *H. buchwaldii* but partially decumbent, with a shorter corolla tube and linear stigma.

NAME ***Harveya andongensis*** Hiern
A species recognised by its flowers with deep yellow, brown-striped tube, blood-red limb and yellow throat, borne in very lax 2–4-flowered cyme; tube 3.8–5.0 cm long, 0.3 cm broad at base, 1.2–1.48 cm broad at apex; lobes 1.6–2.5 cm across; style exserted. Recorded from south-western Africa in Angola, but host and ecology unknown.

NAME ***Harveya buchwaldii*** Engl.
A poorly known species with white or dark purple flowers in racemes to 40 cm; corolla with rounded lobes, clavate stigma. Host unknown; reported from East Africa in eastern Tanzania, growing 900–1,150 m above sea level.

NAME ***Harveya huillensis*** Hiern
A poorly known, small plant 6–16 cm with 1–4 flowers. Calyx glandular-pilose. Corolla glandular-puberulous outside, with obovate lobes; tube narrowly tubular below, inflated above calyx tube, rose-pink to magenta or white with a whitish tube and a yellow or cream throat. Host range and ecology unknown; reported from Angola, east through Zambia and Malawi to Tanzania and Mozambique.

NAME ***Harveya kenyensis*** Hepper
A small, poorly known species to 3–5 cm with glabrous stems and large flowers in racemes. Calyx tubular, short-toothed, 1.1–1.2 cm long, teeth 0.2–0.3 cm. Corolla tube 4–5 cm long, densely pubescent, white with yellow throat, expanding beyond calyx; lobes purple and rounded; style club-shaped, scarcely exserted; filaments glabrous, sterile anther locule long-acuminate. Parasitic on tree roots (species unknown) in dry forests in Kenya, at 30–450 m.

NAME ***Harveya liebuschiana*** Engl. ex Skan
A small, poorly known species to 6–15 cm with clustered, ascending, simple, almost glabrous stems and flowers large relative to the plant, borne in short terminal racemes; scales lanceolate to broadly ovate, 1.5–2.2 cm long, 0.5–1.2 cm broad, ± acute, 1–3-veined. Calyx tubular-campanulate, 2 cm long with unequal, triangular lobes. Corolla with tube narrowed at base and funnel-shaped and curved; lobes spreading to reflexed, lobes subacute, short and broad 3.8–6.8 cm long; white with yellow in the throat. Stigma suborbicular. Filaments inserted 2 cm from the base. Recorded from Uganda and Tanzania; host and ecology unknown.

NAME ***Harveya obtusifolia*** (Benth.) Vatke
A green (hemi-parasitic) plant, 10–60 cm, with erect, simple stems and flowers solitary in upper axils of leafy bracts; bracts elliptic to elliptic-lanceolate, often longer than calyx and corolla; glandular-pubescent. Calyx variably glandular-pilose with long, unequal, lanceolate-triangular, acute lobes. Corolla 2–3 cm long; lobes spreading; tube yellowish, flushed pink; lobes white, reflexed; throat greenish-yellow. Filaments pilose at the apex. Recorded from Saudi Arabia, Yemen and Madagascar in thickets, bushland, fields and orchards, at 350–1,800 m.

NAME ***Harveya tanzanica*** Hepper
A small, poorly known species with lax racemes of 4–10 alternate or opposite flowers. 20–65 cm high; stem, scales, bracts and calyces all dull red to dusky purple; stem rarely branching, glabrous-puberulent below becoming villose towards the apex, 0.3–0.5 cm in diameter, terete or obtusely angled in cross-section. Calyx campanulate at base, cylindrical above. Lobes narrowly triangular, deltoid. Corolla tube narrow below, inflating abruptly beyond the calyx, curved. Lobes imbricate, concave, orbicular to semi-orbicular; colour variable, from creamy white to pale pink or yellow. Stamens inserted at the point of corolla inflation; filaments puberulent-pubescent, anthers glabrous. Host unknown, reported from eastern Tanzania in coastal regions.

NAME ***Harveya thonneri*** De Wild. & T.Durand
A poorly known species, 10–20 cm, with few flowers. Calyx tubular with oblong, triangular lobes, 1.9 cm long. Corolla funnel-shaped, 3.2 cm long, pale yellow, darker yellow in the throat. Stigma capitate. Reported from the Democratic Republic of the Congo; host and ecology unknown.

NAME ***Harveya versicolor*** Engl.
A poorly known species, 15–30 cm, with simple, rather slender, sparingly to densely glandular-pilose stems with few flowers. Calyx glandular-pilose with lobes narrowly triangular, up to 1.0 cm long, the margins glandular-pilose. Corolla tube white externally, yellow within, 5.0–6.0 cm long, 0.4–1.0 cm in diameter; lobes purple, rounded, 1.0–2.0 cm long and wide. Stigma oblong. Stamens with glandular-pilose filaments. Recorded from Tanzania; host and ecology unknown.

NAME ***Harveya vestita*** Hiern
A poorly known species with simple, erect, glandular-puberulent, leafy stems; racemes few-flowered. Calyx campanulate, with lanceolate, obtuse lobes, as long as their tube; 1.9 cm. Corolla funnel-shaped-tubular, veined, the tube broader at the throat, slightly curved; limb concave with rounded lobes; 3.1–3.8 cm long, 0.5 cm wide at base. Throat 1.27 × 1.27 cm. Lobes 1 cm wide. Colour unknown. Stamens with minutely glandular (not hirsute), filiform filaments exceeding the anthers; anthers hairy at top, beaked at base. Recorded from the Cape provinces of South Africa but hosts and ecology unknown. Apparently distinguished by the anther indumentum.

NAME ***Hyobanche atropurpurea*** Bolus

Habit: rhizomatous with one to several long rhizomes; plant glandular-puberulous; dark purple nearly throughout; inflorescence short, dense, subterminal. **Height**: 3–5 cm. **Stem**: short; scales ovate to oblong. **Calyx**: veined, as long as the bracteoles, deeply 5-lobed; segments equal, linear-lanceolate, acuminate, 3 times as long as the tube. Floral bracts oblong or tongue-shaped, obtuse, patent at the apex, 4–5 mm long. Bracteoles 2, linear, acuminate, erecto-patent, 4–5 mm. Upper part of the flower and bracts of a deep claret-red, shading into rose-pink, with yellow at the base. **Corolla**: ca. 1 cm long, blackish, dusky purple, cucullate, veined, shortly tubular, sparingly puberulous outside, glabrous within; mouth obovate, widely open, oblique; lobes triangular, the frontmost subulate. **Carpel**: stigma pinkish, yellowish; style shortly bilamellate and thickened at the apex, glabrous. **Stamens**: not exceeding the corolla. **Host**: unknown. **Ecology and distribution**: occurs in the Western Cape Province of South Africa in rocky habitats and sandstone fynbos at higher elevation. Flowering usually (September) October–November (December). Pollinated by Cape rock elephant-shrews. **Key diagnostics**: distinguished by its blackish, dusky purple (extremely the whole plant could be yellow) colouration and acuminate (rather than obtuse) bracteoles; corolla somewhat hooded. **Similar species**: none.

Hyobanche atropurpurea
South Africa
(left, page 204) photos: Corinne Merry
(right, page 205) photos: Ismail Ebrahim

NAME ***Hyobanche robusta*** Schönland

Habit: robust; rhizomatous with one to several long rhizomes (1–2 m long). **Height**: inflorescence up to 15 cm. **Stem**: up to 23 cm long, 2–5 cm thick, fleshy. Scales squamiform, densely imbricate; lower 5–7 mm long, orbicular-ovate, abruptly acute, sub-glabrous, brownish-ferruginous; uppermost gradually longer, up to 3 cm long and 1–2 cm wide, oblong or elongate-oblong, acute or obtuse, villose-tomentose, dorsally ferruginous, the others white. **Calyx**: floral bracts 2–3 cm long, 1–1.2 cm wide, oblong, obtuse, villous-tomentose. Bracteoles 2–3 cm long, 1.5–2 mm wide, linear, acute, villous-tomentose, white. **Corolla**: large, above 7–8 mm across, procumbent, apex subacute, mouth oblique, 2 cm long, unidentate, externally villous-tomentose, glabrous within, base yellowish, apically ferrugineous. **Carpel**: stigma whitish or yellowish; ovary globose, glabrous; style apex curved, flattened-sub-clavate. **Stamens**: filaments barely exserted and basally pubescent; anthers yellow-brown. **Host**: unknown (possibly various coastal plants). **Ecology and distribution**: coastal dunes in the Eastern Cape Province (south-east coast between Knysna and East London). **Key diagnostics**: rather robust, congested, many-flowered spicate, dull pink inflorescences; flowers patent; stigma whitish or yellowish, exserted. **Similar species**: similar to *H. sanguinea*, but more erect, with a sepal posterior free, corolla longer, with sharper apex and larger mouth.

Hyobanche robusta
South Africa (right)
photo: Sue Marsden

NAME ***Hyobanche rubra*** N.E.Br.
SYNONYM ***Hyobanche glabrata*** Hiern

Habit: a small, clump-forming plant with subterranean stems and densely corymbose-spicate inflorescences at ground-level. **Height**: 5 cm. **Stem**: fleshy, almost woody at the base; scales oblong, obtuse, 3 cm long, ciliolate, otherwise glabrous. **Calyx**: acutely 5-lobed, sparsely pilose, 1 lobe linear and free, the rest linear-falcate and connate to the middle; 3.5 cm long. **Corolla**: sub-erect, tubular, sub-cylindrical, scarcely curved; exterior pubescent outside, glabrous within; 5 cm long; mouth very oblique with a very short tooth at the base; corolla pinkish-red to orange. **Carpel**: stigma prominent, slightly bilobed, dorsally compressed, pale yellow; style subequalling the corolla, curved at the apex, glabrous. **Stamens**: glabrous. **Host**: shrubby Asteraceae (e.g. *Pteronia incana*). **Ecology and distribution**: Northern and Western Capes of South Africa on dunes and sand flats. **Key diagnostics**: small clusters of long-tubular, sub-erect, reddish flowers appearing at ground-level with no visible stem. **Similar species**: *Hyobanche glabrata* and *H. calvescens*, once widely recognised, are now regarded to be synonyms of *H. rubra*.

Hyobanche rubra
South Africa (right)
photo: Mayur Prag

NAME ***Hyobanche sanguinea*** L.
SYNONYMS ***Haematobanche sanguinea*** (L.) C.Presl; ***Hyobanche coccinea*** Steud.

Habit: a small, erect plant with dense, oblong to pyramidal spicate inflorescences. **Height**: to 15 cm. **Stem**: fleshy, almost woody at the base. Bracts oblong, imbricate, obtuse, 3 cm long, ciliolate, otherwise glabrous. **Calyx**: acutely 5-lobed, sparsely pilose, 1 lobe linear and free, the rest linear-falcate and connate to the middle; 3.5 cm long. **Corolla**: sub-erect, tubular-cylindrical, curved, hooded; densely glandular-pubescent throughout; 5 cm long; corolla bright pinkish-red. **Carpel**: stigma prominently exserted, pale yellow; style curved at the apex, glabrous. **Stamens**: dehiscent. **Host**: *Aspalathus* (Fabaceae), *Hermannia* (Malvaceae), *Metalasia* (Asteraceae) and *Protea* (Proteaceae). **Ecology and distribution**: widespread in Lesotho and South Africa (Eastern Cape, Northern Cape, Orange Free State, Western Cape; Swaziland), on dunes and sand flats. **Key diagnostics**: densely glandular-pubescent spikes of sessile, bright pink-red flowers borne in oblong to pyramidal spikes. **Similar species**: *Hyobanche thinophila* is similar but with flowers glabrous within.

Hyobanche sanguinea South Africa (above, right) photos: Sebastian Hatt

NAME ***Hyobanche thinophila*** A.Wolfe

Habit: rhizomatous with one to several yellowish subterranean rhizomes (exceeding a metre in length) extending deep into sand, with numerous secondary haustoria forming from scales; inflorescence densely flowered, ovoid to slightly oblong, 8.3–15.4 × 3.8–6.7 cm; flowers sessile to subsessile, tubular with 2 lateral bracteoles, subtended by a single bract on the abaxial side. **Height**: inflorescence up 15 cm. **Stem**: scales ovate to oblong, 5.0–13.0 × 4.0–10.0 mm, apex acute to slightly obtuse, upper margins ciliate with short multicellular trichomes, abaxial surfaces of above-ground scales puberulent near apex, scales mostly glabrous below-ground. **Calyx**: 3.8–5.1 cm, 4 of 5 sepals fused to ¼–½ their length, the abaxial sepal free or connected to other sepals near the calyx base, sepal lobes lanceolate to broadly linear, acute, pilose. Floral bracts 1.9–3.4 × 0.7–1.4 cm, oblong and keeled adaxially, acute to slightly obtuse, becoming pubescent to pilose. Bracteoles 2.7–3.9 × 0.2–0.5 cm, lanceolate to broadly linear, acute to acuminate, the upper half pilose. **Corolla**: 5.6–6.2 × 1.0–1.4 cm, tubular or slightly curved, slightly inflated above the tube constriction; 1.1–1.8 × 0.5–0.9 cm; light pink, densely pilose with 2 minute lapel-like lateral lobes and a small denticle at the base of the corolla mouth; lateral lobes slightly to fully reflexed. **Carpel**: pistil recurved ca. 90° below stigma, stigma exserted, clavate to nearly capitate, slightly bilobed dorsiventrally. **Stamens**: 4, inserted in the corolla at the point of constriction, 3.4–4.2 cm; filaments pubescent to tomentose; anthers calceiform, one-celled, with a small knob or spur on the proximal end from vestigial theca, apical pore elongated and slightly recurved to slightly exserted at maturity. **Host**: *Metalasia* and *Chrysanthemoides* (Asteraceae) and *Passerina* (Thymelaeaceae). **Ecology and distribution**: sand dune and fynbos along the coast of South Africa, Western Cape, from Struisbaai in the south to Yzerfontein in the west. Flowers in late winter to early spring (August–September). Name 'inkblom' refers to the use of the plant (as others *Hyobanche*) for making ink. **Key diagnostics**: ovoid to slightly elongated inflorescences; flowers are light pink, tubular and non-galeate, and slightly curved to straight; a ring of long multicellular hairs at the level of stamen insertion. **Similar species**: *Hyobanche sanguinea* differs in having a compact to oblong inflorescence; flowers scarlet to deep red, galeate and strongly curved; short glandular hairs scattered throughout the interior surface of the corolla. Phylogenetically more closely related to *H. robusta* than *H. sanguinea*.

Additional species

NAME ***Hyobanche hanekomii*** A.Wolfe

Similar to *H. sanguinea* but with corolla deep magenta to magenta red, semi-galeate (hooded), strongly inflated above tube, corolla length 1.5–2.0 times length of calyx. Possibly parasitic on *Elytropappus* (Asteraceae), *Passerina* (Thymelaeaceae), *Phylica* (Rhamnaceae) and *Restio* (Restionaceae) in fynbos. Western Cape, South Africa. Other similar species include: ***Hyobanche barklyi*** N.E. Br. in Namibia, South Africa: north-west Northern Cape, and ***H. fulleri*** E. Phillips on the coast of KwaZulu-Natal from Durban to Port Shepstone in South Africa.

Hyobanche thinophila
South Africa (right)
photo: Gigi Laidler

NAME ***Kopsiopsis strobilacea*** (A.Gray) Beck
SYNONYM ***Boschniakia strobilacea*** A.Gray

Habit: robust, cone-like, ± compact inflorescences with numerous flowers from ground-level. **Height**: 10–30 cm. **Stem**: robust, dark maroon or purple (sometimes the whole plant is yellow); scales imbricate, spathulate or oblanceolate. **Calyx**: with 0–5 tapered lobes. Floral bracts upward-swept, yellowish. **Corolla**: narrowly tubular; upper lip lobes fused; lower lip lobes divergent; tube 15–20 mm, dark brownish-purple externally, distally pale pink; lobes pinkish with whitish, glabrous margins. **Carpel**: stigma capitate, yellowish; style as long as, or shorter than, stamens. **Stamens**: with a clump of hair at the filament base and villous anthers. **Host**: *Arbutus menziesii* and *Arctostaphylos* spp. (Ericaceae). **Ecology and distribution**: temperate woods and chaparral from Oregon to Baja California, and Mexico to elevation of 1,830 m. Flowering April–July. **Key diagnostics**: dark brownish-purple, cone-like inflorescences. **Similar species**: ***Kopsiopsis hookeri*** (Walp.) Govaerts is similar but is a much smaller plant and distinguished by its lanceolate to obovate (rather than ovate to spoon-shaped) bracts, smaller flowers, 10–15 mm, and is parasitic on *Gaultheria shallon*, and also reported on *Arctostaphylos*, *Vaccinium* and *Arbutus* (Ericaceae); this species occurs from western Canada (Queen Charlotte Islands) to California.

Kopsiopsis hookeri Oregon, USA (above left, above right) photos: Gerald D. Carr, Canada (page 216, 217) photos: Karen L. Barry
Kopsiopsis strobilacea (right) photo: Paul Hankamp

NAME ***Lathraea clandestina*** L.

SYNONYMS ***Clandestina pallidiflora*** Martrin-Donos; ***Clandestina penduliflora*** Lam.; ***Clandestina rectiflora*** Lam.; ***Clandestina clandestina*** (L.) Huth

Habit: rhizomatous, glabrous perennial, extensively patch-forming above-ground made up of congested corymbose raceme or spicate inflorescences, each with 4–8 flowers; aerial stems short to absent. **Height**: 5–8 cm. **Stem**: creeping, lacking conspicuous aerial shoots: a substantive, cream-coloured branching rhizome system with succulent scales; scales fleshy, reniform, amplexicaul, alternate or opposite, each with a lacuna with capitate hairs and dome-shaped glands. **Calyx**: campanulate, with 4 short equal triangular lobes above and 4 prominent nerves, pale lilac; 1.8–2.0 cm. **Corolla**: tubular, sub-erect; upper lip hooded, lower lip 3-lobed, exceeded by the upper lip. Tube 4.0–5.0 cm, purple-violet, occasionally white or pink. **Carpel**: stigma disc-shaped, pale; style terminal, exerted, slender down-curved at tip, projecting from the tip of the corolla. **Stamens**: exserted, with anthers clustered. **Host**: various tree families including Salicaceae and Sapindaceae (occasionally reported on Betulaceae, Buxaceae, Juglandaceae, Cupressaceae, Ericaceae, Taxaceae and Gunneraceae). **Ecology and distribution**: native to western and southern Europe, from Belgium, France to northern Spain and central and southern Italy, in damp, shaded woods; introduced to other countries, e.g. Austria, Germany, Great Britain, Ireland, mainly in parks and old gardens, and frequently naturalised. Flowering February–July, depending on location and elevation. **Key diagnostics**: extensive patches of purple-flowered inflorescences. **Similar species**: none.

Lathraea clandestina UK (above, right, page 220, 221)

NAME ***Lathraea japonica*** Maxim. ex Franch. & Sav.
SYNONYMS ***Clandestina japonica*** Miq.; ***Lathraea chinfushanica*** Hu & Tang

Habit: erect, many-flowered, loosely narrowly cylindrical inflorescences. **Height**: 20–30(–35) cm. **Stem**: sub-glabrous below, with yellow-brown glandular hairs above; scales white, diamond-shaped, broadly ovate, or semicircular at the stem base. **Calyx**: tubular-campanulate, with stalked glandular hairs on exterior, white to pale rosy or rarely purple, often with dark purple dot at splitting point of calyx lobes, 7–9 mm long, mostly equally 4-lobed. **Corolla**: narrowly tubular, 10–17 mm long, mostly glabrous; upper lip hooded, slightly emarginate, 4–5 mm long; lower lip equally 3-lobed, broad, crispate towards the margins, 2.5–3 mm long; upper lip slightly longer than the lower; corolla more than twice as long as calyx when mature; tube white to pale rosy or rarely purple. **Carpel**: stigma pale, capitate, scarcely bilobed; style filiform. **Stamens**: scarcely exserted beyond the upper lip; filaments filiform, papillose-verrucose. **Host**: various possible trees, potentially Fagaceae, Betulaceae and Salicaceae. **Ecology and distribution**: deciduous forests in central and south China, South Korea, central and south Japan, at (70–)1,200–3,000 m. Flowers March–June. **Key diagnostics**: early flowering, loosely narrowly cylindrical and many-flowered inflorescences. **Similar species**: *Lathraea squamaria* is similar but has 1-sided inflorescences.

Lathraea japonica
Japan (right)
photo: Ben Jones

NAME ***Lathraea rhodopea*** Dingler

Habit: erect, many-flowered, dense, cylindrical inflorescences. **Height**: to 60 cm. **Stem**: stout, yellow, with reniform to ovate clasping whitish bracts. **Calyx**: shortly tubular-campanulate (much exceeded by the corolla), with stalked glandular hairs on exterior, purplish in flower, yellow in fruit, 5–10 mm long, unevenly 4-lobed. **Corolla**: narrowly tubular, 9–12 mm long, glabrous; upper lip hooded, 3 mm long; lower lip shortly 3-lobed, 2 mm long, with lateral lobes broadly ovate, entire, with middle lobe apically truncate and tridentate, upper lip slightly longer than lower lip. **Carpel**: stigma capitate, pale; style filiform. **Stamens**: exserted up to 2 mm beyond corolla; filament filiform, 4–6 mm long, papillose-verrucose. **Host**: tree roots, most notably *Corylus* (Betulaceae) and *Salix* (Salicaceae). **Ecology and distribution**: native to Balkan Peninsula, from southern Bulgaria to north-east Greece in temperate deciduous woods and by ravines or creeks, at 450–1,600 m. Flowers March–May. **Key diagnostics**: dense, cylindrical inflorescence of pale pink flowers. **Similar species**: somewhat similar to *L. squamaria*, which is far smaller and has 1-sided inflorescences with less yellow pigmentation.

Lathraea rhodopea
Macedonia, Greece
(left, right, page 226, 227)

NAME ***Lathraea squamaria*** L.
SYNONYMS ***Clandestina squamaria*** (L.) Legrand; ***Squamaria orobanche*** Scop.

Habit: erect, dense, 1-sided inflorescences. **Height**: 8–30 cm. **Stem**: fleshy, aerial, erect stems borne from much-branching, pale rhizomes. **Calyx**: shortly tubular-campanulate, with stalked glandular hairs to 2 mm long on exterior, pale pinkish-lilac to pink-purple, 8–12 mm long, almost evenly 4-lobed. **Corolla**: narrowly tubular, 12–17(–20) mm long, glabrous, bilabiate; upper lip hooded, 4–6 mm long; lower lip shortly 3-lobed, 3–5 mm long; upper lip slightly longer than lower lip; tube pale pink. **Carpel**: stigma pale, capitate, scarcely bilobed; style filiform. **Stamens**: exserted up to 2 mm beyond corolla; filament filiform, 5–6 mm long, papillose-verrucose. **Host**: various trees, especially *Corylus* (Betulaceae) and *Ulmus* (Ulmaceae). **Ecology and distribution**: widespread across Europe to north Iran, in temperate, broad-leaved forests and hedgerows, at elevation 0–3,000 m; early-flowering plant (March–early May (July) depending on altitude). **Key diagnostics**: early-flowering, pale pink, 1-sided spikes. **Similar species**: somewhat similar to *L. rhodopea,* which lacks the 1-sided spikes, is far more robust and turns yellowish with age.

Additional species

NAME ***Lathraea purpurea*** H.A.Cummins ex King
A pale mauve, densely white-hairy species with flowers borne at ground-level, distinguished by its unlobed calyx with an undulate margin, corolla 19–25 mm long, and its marked host preference for bamboo (*Yushania niitakayamensis*) and other related Poaceae, in India (eastern Himalayas) and Taiwan, usually at elevation 3,000–4,000 m. Flowers July–September. Unpublished molecular work suggests this species in fact belongs to its own genus allied to *Pedicularis*.

Lathraea squamaria
UK (left, right)

NAME ***Mannagettaea hummelii*** Harry Sm.

Habit: clump-forming, patch-forming perennial with congested corymbs of flowers. **Height**: 3–5 cm tall. **Stem**: thick, yellowish; aerial portion absent; scales broadly ovate-triangular, 6–7 × 6–7 mm, glabrous, apex obtuse. **Calyx**: tubular with triangular, subequal lobes 3–5 × 2–4 mm; tube 1.5–1.7 cm. **Corolla**: tube slightly curved, densely villous at apex and adaxially; upper lip entire to 2-lobed, 5–8 mm; lower lip 3-lobed; lobes linear, 4–5 × 1–2 mm; tube 2.2–2.5(–3) cm; corolla pale yellowish below, dark dull purple distally. **Carpel**: stigma sub-globose; style glabrescent, 8–10 mm. **Stamens**: filaments 1–1.5 cm, densely woolly beneath; anthers ovoid. **Host**: *Caragana* (Fabaceae) and *Salix* (Salicaceae). **Ecology and distribution**: Siberia to China (Qinghai, Gansu, and Xizang), in spruce forests, understory vegetation and among shrubs on hillside 3,200–3,700(–4,460) m. Flowering July–August. **Key diagnostics**: patch-forming; congested corymbs of purplish flowers. **Similar species**: *Mannagettaea labiata* Harry Sm., in China (Sichuan), is a similar species, with spikes to 11 cm and larger flowers to 6 cm.

Mannagettaea hummelii
Illustrated (right)
Buryatia, Russia (page 232, 233)
photos: Daba G. Chimitov

NAME ***Orobanche alba*** Stephan ex Willd.
SYNONYMS ***Orobanche epithymum*** DC.; ***Orobanche rubra*** Sm.; ***Orobanche thymi-serpilli*** Vaucher ex Duby

Habit: rather short, few-flowered inflorescences, rarely taller plants with dense inflorescences. **Height**: 8–30(–60) cm. **Stem**: simple, stout, with few scales; yellow to dark orange-red. **Calyx**: usually entire, occasionally bidentate, about equalling the corolla tube; 8–16 mm. Floral bracts 12–25 mm, lanceolate, acuminate, glandular, slightly shorter or equalling the length of the corolla. **Corolla**: glandular-pubescent especially distally, erecto-patent, slightly curved, more strongly so at the basal and distal ends; tube campanulate-cylindrical; (12–)15–25(–30) mm long, upper lip entire or 2-lobed with somewhat spreading margins; lower lip glandular-ciliate, the 3 lobes about equal, the middle just the largest; 8–16 mm; rather shiny pale cream, usually strongly suffused dull reddish-purple. **Carpel**: stigma lobes usually reddish or purplish, rarely orange, yellow, the lobes touching. **Stamens**: inserted 1–3 mm above the base of the corolla tube; filaments slightly hairy below, glandular above. **Host**: various Lamiaceae, especially *Thymus* spp., *Salvia* spp., *Clinopodium vulgare*, *Origanum vulgare*, *Ziziphora* spp., *Micromeria* spp. and *Satureja* spp. **Distribution and ecology**: rocky cliffs and scree in the north of the range; xerothermic grasslands, maquis, scrub and mountain pastures further south; widespread and fairly common across Europe to the central Himalaya, south-east Tibet to China (north-west Sichuan), and northern Africa. Flowering June–August depending on location and elevation. **Key diagnostics**: polymorphic taxon, but has fragrant flowers, broadly campanulate-cylindrical corolla with distinct dark glands and (usually) subentire to entire calyx-teeth. **Similar species**: similar to *O. raddeana*, which is restricted to Campanulaceae in the Caucasus region; *O. teucrii* is also found on Lamiaceae (*Teucrium*) and is somewhat similar but has divided calyx segments, and a narrowly infundibuliform corolla without dark glands.

Infraspecific taxa (selected)

Subsp. *alba* is the widespread form of the species, has smaller and less numerous flowers and typically grows on *Thymus*. **f. *rubra*** is a dark orange-red form that occurs in rocky, mainly coastal habitats in the UK.

Subsp. *cuprea* has large, conspicuous pale yellow folds between lobes of the lower lip and an absence of dark glands or darker hairs on the exterior of the corolla. Cyprus and southern Turkey.

Subsp. *major* has larger, light-coloured, usually numerous flowers with widely open throats, longer bracts, sepals and scales, and is parasitic on *Salvia*. Distribution poorly known; mainly central and south to south-east Europe.

Subsp. *xanthostigma* has free calyx segments and bright yellow stigma. **f. *typica*** has corolla yellowish-red towards the margins; red-violet veined; **f. *sineglandulosa*** has entirely yellow flowers. Caucasus and probably elsewhere.

Orobanche alba subsp. *alba* f. *rubra* Northern Ireland, UK (page 236)
Orobanche alba Greece (page 237)
Orobanche alba subsp. *cuprea* Cyprus (page 238) photo: Christodoulos Makris (*Flora of Cyprus*)
Orobanche alba subsp. *major* Poland (page 239)

O. alba subsp. *xanthostigma*

O. alba subsp. *alba* f. *ruba*

O. alba subsp. *major*

O. alba subsp. *cuprea*

Orobanche alba subsp. *xanthostigma* Armenia

NAME ***Orobanche almeriensis*** A.Pujadas

Habit: rather slender with lax flowers on most of, or the upper ½–⅔, of the stem. **Height**: 12–20 cm. **Stem**: initially robust, later slender; reddish-purple; scales rather long, not recurved; triangular, glabrous and upper longer, hairy and turning dark brown. **Calyx**: teeth fused, uneven, 9–12 mm; floral bracts 13–16 mm, equalling, shorter or just exceeding the corolla. **Corolla**: ± erect, straight; upper lip 2-lobed; lower lip with subequal lobes, all notched; tube cream, flushed pinkish-purple; lobes cream with darker veins; corolla 1.7–2.0 cm; tube densely glandular-pubescent, basally glabrescent. **Carpel**: stigma dark purple; lobes just touching; style sparsely glandular. **Stamens**: filaments inserted 2–3 mm above the base of the corolla, sparsely hairy below, glabrous above. **Host**: *Andryala ragusina* (Asteraceae); possibly other plants. **Ecology and distribution**: Andalusia, Spain; disturbed habitats such as roadsides. Flowering mid-April and May. **Similar species**: *Orobanche amethystea* is very similar but has a different host range, bracts much exceeding the corolla, calyx segments free; filaments inserted 3–4 mm above the corolla base. *Orobanche andryalae* also grows on *Andryala* but has a markedly cernuous corolla, has more yellowish-orange pigmentation, and is restricted to volcanic habitats in the Canary Islands.

Orobanche almeriensis Spain (above) photo: Enrique Triano, (right, pages 242, 243)

NAME ***Orobanche alsatica*** Kirschl.

SYNONYMS ***Orobanche minor*** **subsp.** ***alsatica*** (Kirschl.) Nyman; ***Orobanche cervariae*** Kirschl. ex Suard

Habit: usually stout; inflorescence cylindrical to ovate, usually dense and many-flowered. **Height**: (25–)40–70 cm. **Stem**: stout, erect, yellow, yellowish-brown, pinkish, glandular-pubescent; scales lanceolate to triangular. **Calyx**: segments usually bidentate, free or connate in front, about a ⅓ or ½ of the length of the corolla, with lanceolate-subulate teeth, glandular-pubescent, 7–12 mm. Floral bract about ⅔ of the length of the corolla, lanceolate, glandular-pubescent. **Corolla**: broadly tubular, with regularly and strongly curved dorsal line, 20–23(–25) mm. Yellow, yellowish, yellowish-brown, yellowish-pink with darker veins. Upper lip almost entire or slightly emarginated, with broad, porrect, erect lobes; lower lip 3-lobed, lobes irregularly crenate and deflexed. **Carpel**: stigma bilobed, yellow; style glandular-pubescent. **Stamens**: filaments inserted 4–7 mm, pubescent at the base and sparsely glandular-pubescent or glabrous near the anthers; anthers usually pubescent at the line of fusion. **Host**: *Peucedanum* spp., including *P. cervaria* and *P. alsaticum* (Apiaceae). **Ecology and distribution**: xerothermic grasslands, open forests, dry meadows, rocky slopes. Flowering usually mid-June–end July. France to southern and central Europe; occurrence further east to be confirmed. **Key diagnostics**: usually stout, tall plant, yellow, yellowish-brown or pinkish, inflorescence usually cylindrical, dense and many-flowered; corolla broadly tubular, with regularly and strongly curved dorsal line, relatively high filaments insertion level. **Similar species**: *Orobanche alsatica* is morphologically similar and closely related to *O. bartlingii*, but differs in the size of the corolla, which is larger in the former (20–25 mm rather than 12–18 mm), shape of the corolla (regularly and strongly curved dorsal line rather than evenly curved at the base), style that is glandular-pubescent rather than glabrous or rarely glandular-pubescent, and stamens that are inserted 4–7 mm vs. 1–3 mm above base of the corolla tube; it also has a different host range (*Peucedanum* rather than *Seseli*).

Orobanche alsatica
Poland (right)

NAME ***Orobanche amethystea*** Thuill.
SYNONYMS ***Orobanche amethystea* var. *typica*** Beck; ***Orobanche amethystina*** Rchb.; ***Orobanche attica*** Reut.; ***Orobanche eryngii*** Duby

Habit: inflorescences with numerous flowers on most of, or the upper ½–⅔, of the stem, congested above; spikes with a somewhat regular flower arrangement. **Height**: 15–50 cm. **Stem**: red-brown or violet, rarely yellowish, glandular-pubescent, with numerous scales at the base and usually with fewer above; lower scales erect to obliquely patent, triangular to broadly lanceolate and glabrous; upper scales lanceolate, glandular-pubescent, obliquely patent. **Calyx**: segments free, unequal; 14–17 mm, markedly exceeding the corolla. **Corolla**: sharply inflected at the base, rather straight along the dorsal line then abruptly geniculate at the tip; 16–25 mm, cream with purple veins; glandular-pubescent, basally glabrescent or glabrous. **Carpel**: stigma pink to violet (sometimes orange-yellow or yellow); style sparsely glandular-pubescent. **Stamens**: filaments hairy below, glabrous above; inserted 3–4 mm above the corolla base. **Host**: usually *Eryngium* spp. (Apiaceae) but occasionally other herbs such as *Daucus*, *Ballota* and various Asteraceae. **Ecology and distribution**: western and central Mediterranean and Europe on sea cliffs, garrigue, sand dunes and disturbed habitats. Flowering end April and May (June). **Key diagnostics**: rather straight then geniculate corolla dorsal line, often (but not always) with strong purple pigmentation; calyx segments (1–)3–4-veined. **Similar species**: *Orobanche crenata* can be similar in stature and general appearance, but has white, campanulate, fragrant flowers; *O. almeriensis* has shorter floral bracts; closely-related *O. hederae* can be similar in general appearance but has a yellow stigma, a narrower corolla and different ecology; robust forms of *O. minor* can share a resemblance (and share host species) but have fewer, smaller flowers with smoothly curved dorsal line, filaments glabrescent or sparsely hairy below, inserted 2–3 mm above the corolla base, shorter bracts and 1-veined calyx segments.

Infraspecific taxa (selected)

Subsp. *amethystea* has 1-veined calyx segments that are apically filiform and a deeply 2-lobed upper corolla lip. Parasitic on Apiaceae (and other herbs listed above). Throughout the species range.

Subsp. *castellana* has 3–4-veined calyx segments that are not apically filiform and entire to 2-lobed upper corolla lip. Parasitic on *Digitalis* (Plantaginaceae). South-west Europe.

Orobanche amethystea
Albania (left)
Spain (right, page 249) photos: Francisco Serrano Ortiz
Germany (page 248) photo: Michael Pollich

NAME ***Orobanche amoena*** C.A.Mey.
SYNONYM ***Orobanche colossea*** (Beck) Novopokr.

Habit: usually somewhat robust; inflorescence ovoid or short-cylindrical, rather dense, rarely lax, shorter or equalling its stem. **Height**: 15–40 cm. **Stem**: thickened at the base with ovate-lanceolate to lanceolate scales, 12–18 mm long; yellowish, violet, reddish, blue-grey, glandular-pubescent. **Calyx**: segments 7–15 mm long, entire or bidentate, teeth narrowly lanceolate and subulate at the tip. Floral bracts ovate-lanceolate, 7–18 mm long, shorter than ½ of the corolla. **Corolla**: 20–35(–40) mm long, usually erect to erecto-patent, tubular to tubular-campanulate or infundibuliform, gradually but considerably broadened, sometimes slightly curved from constriction above the ovary, dark blue or violet (sometime whitish-blue and violet only on the lips), usually sub-glabrous. Upper lip 2-lobed, lower lip slightly longer than upper, lobes broad, oblong with margin irregularly dentate or equal. **Carpel**: capsule oblong; stigma usually 2-lobed, whitish; style sub-glabrous. **Stamens**: filaments pubescent at the base, inserted 6–11 mm from the corolla base; anthers pubescent along sutures. **Host**: Asteraceae, usually *Artemisia* spp. **Ecology and distribution**: steppe, rocky slopes. Flowering June–July. European Russia to Transcaspian, central Asia, Mongolia, China, north-west Himalayas to Pakistan. **Key diagnostics**: corolla with narrow tube and broad dark blue or violet lobes. **Similar species**: polymorphic in size and corolla colour; smaller specimens may be confused with *O. cumana*, which also occurs on the steppe and on *Artemisia*; larger specimens (f. *colossea*) may resemble *O. gigantea*. Similar to the poorly known *O. hansii* A.Kern. (transitional species between *O. amoena* and *O. cernua* described from India, on *Artemisia*), in form, colour and size of corolla and *O. sintenisii* Beck described from Turkey may also be similar. Probably similar also to *O. sulphurea* Gontsch. (described from Tajikistan), which may in fact be a hybrid of *O. amoena* and *O. kotschyi*, as well as *O. cistanchoides* Beck (type from Iran). Probably synonymous with *O. stocksii* Boiss. (type from Afghanistan, reported on *Eryngium* and *Cousinia*), which differs in having a campanulate corolla that is enlarged above the filament insertion and narrower below, subtubular, and the entire corolla enlarged from the base upwards. Each of the species cited above require further investigation. *Orobanche clarkei* Hook.f. (not shown) is similar and also parasitic on *Artemisia*, but puberulous (not subglabrous), with more numerous flowers and less campanulate corolla. Reported from Afghanistan, China, India, Kazakhstan, Pakistan and Tajikistan.

Orobanche amoena
Kazakhstan (left, right)
photos: Alexander Fateryga

NAME ***Orobanche andryalae*** C.J.Thorogood, M.Hernandez Gonzalez, Rumsey & Reyes-Bet.

SYNONYMS ***Orobanche amethystea*** Thuill.; ***Orobanche castellana*** Reut.

Habit: inflorescences lax to rather dense above, with 5–15(–20) flowers on most of, or the upper ¼ to ⅓, of the stem. **Height**: 6–16(–25) cm. **Stem**: glandular-hairy, pale orange to light reddish-brown, scarcely swollen below; subterranean scales broadly triangular, yellow; those above rather sparse, brown, 8–15 mm. **Calyx**: 5–7 mm with segments fused, strongly unequal (rarely entire), not exceeding the corolla tube; floral bracts 10–12 mm, rather shorter than the corolla, broadly triangular, brown, glandular-hairy. **Corolla**: strongly cernuous when mature, remaining so in fruit (rarely geniculate); 10–15 mm, pale yellow with faint reddish veins and scattered glandular hairs; upper lip bilobed; lower lip 3-lobed, the lateral lobes slightly exceeding the central; all lobes minutely toothed. **Carpel**: stigma lobes touching, mid to dark red-orange. **Stamens**: filaments sparsely hairy below, glabrous above; inserted conspicuously (ca. 5 mm) above the corolla base; anthers ± glabrous. **Host**: *Andryala* spp. (Asteraceae): *A. perezii* on Lanzarote and Fuerteventura, and *A. pinnatifida* on Tenerife. **Ecology and distribution**: volcanic slopes and steep sea cliffs in the Canary Islands (Fuerteventura, Lanzarote and Tenerife). **Key diagnostics**: *Orobanche andryalae* can be readily distinguished by its strongly cernuous corolla, pale orange colouration, high filament insertion and distinct host and ecology. **Similar species**: historically confused with *O. amethystea* subsp. *castellana*, which has a different corolla profile (although we have seen geniculate corolla profiles in *O. andryalae*), lower filament insertion, and longer calyx lobes and bracts. The most similar species is *O. calendulae*, but this itself is a conflicted taxon: various floras describe equal calyx segments that are fused at the base, but we note that the calyx segments in the holotype from Algeria are bifid but somewhat unequal. *Orobanche calendulae* has a host preference for *Calendula* spp. and smoothly curved corolla dorsal line and filaments that are hairy below, glabrescent above and inserted 2–4 mm above the corolla base, which together distinguish this species.

Orobanche andryalae
in cultivation (left)
(Oxford; source: Lanzarote)
Lanzarote, Canary Islands (right)
photo: Alfredo Reyes Betancort

NAME ***Orobanche apuana*** Domina & Soldano

Habit: rather robust, spikes of lax flowers over most of the stem. **Height**: 15–30 cm. **Stem**: erect, curved, reddish-brown or yellow-brown, densely glandular-pubescent, with few scales; lower scales triangular to ovate, sparsely glandular-pubescent; upper scales linear-lanceolate and sparsely glandular-pubescent, brown, erect. **Calyx**: segments entire or bifid, shorter than their tube, upper segment longer than the lower, or subequal. **Corolla**: dorsal line of corolla slightly curved in lower part, almost straight in the middle, and strongly forward-directed and geniculate behind the upper lip, 1.7–2.4 cm long, yellow to pinkish with few purple veins; upper lip almost entire or slightly emarginate with 2 crenate lobes, with glandular hairs; lower lip lobes equal, crenate, deflexed. **Carpel**: stigma yellow, bilobate; style glandular-pubescent. **Stamens**: inserted 0.3–0.5 cm above the base of the corolla; filaments pubescent at the base; anthers pubescent at the line of fusion. **Host**: exclusively parasitic on *Santolina pinnata* (Asteraceae). **Ecology and distribution**: Apuan Alps, central Italy, on the western slope of Mount Tambura on sunny carbonate rocks and dry pastures on lithosol. **Key diagnostics**: beige to pale orange spikes and distinct host range. **Similar species**: somewhat similar to species from Subsection *Minores,* but distinguished by the shape and dimensions of the corolla, calyx morphology and floral bracts.

Orobanche apuana
Italy (right)
photo: Adriano Soldano

NAME ***Orobanche arpica*** **Piwow., Ó.Sánchez & Moreno Mor.**

Habit: slender or slightly robust; inflorescence (10–)12–13(–15) cm, cylindrical, shorter than the remaining stem, usually dense or rarely lax. **Height**: (20–)27–30(–35) cm. **Stem**: pale white-yellow or rarely pale yellow-brown-pink, glandular-pubescent especially in the upper part; scales (15–)20–24(–25) mm, lanceolate. **Calyx**: segments free (distinctly separate), simple, entire or rarely unequally bidentate, with second small teeth; narrow, with teeth acuminate to almost apically filiform, segments with ovate bases; (11–)13–14(–15) mm long, shorter than ½ the corolla tube. Floral bracts (16–)20–23(–26) mm, shorter or rarely equal to the corolla, broad ovate-lanceolate, yellow-pink or pinkish-brown, with dense white or pale yellow glandular hairs. **Corolla**: ± erecto-patent, tubular to narrowly campanulate (slightly widening towards the mouth), strongly inflated above the insertion of the filaments and strongly constricted below; (22–)24–26(–28) mm long, (8–)9–10(–11) mm in diameter in the central part; pale pinkish-yellow, pinkish-white or yellowish-pink, with dirty pale pink veins; upper lip slightly emarginate, with 2 broad lobes, these usually facing downward, more rarely slightly bent backward laterally, sparsely glandular-pubescent also in the inner part; lower lip with 3 ovate, or ± rectangular to spathulate lobes, central lobe slightly larger than lateral lobes, lobes slightly and irregularly dentate on margins and sparsely glandular-pubescent also in the inner part. **Carpel**: stigma lobes elongated-spherical to ovate and ± flattened, yellow or bright yellow; style glandular-pubescent, especially on the upper part. **Stamens**: obliquely inserted at 4–5(–6) mm above the corolla base; filaments densely villous, from base to ± middle part, upper part densely glandular-pubescent; anthers basally and along ¾ of suture distinctly hairy. **Host**: *Psephellus pulcherrimus* (Asteraceae). **Ecology and distribution**: known only from the type locality, in a subalpine meadow, 2,225–2,230 m. Flowering in the first half of July. Probably a Caucasian endemic, described from the Lesser Caucasus in Armenia. **Key diagnostics**: pale yellowish colour of the plant, and the strong narrowing of the corolla at the point of filament insertion; corolla pale ± pinkish-yellow, pinkish-white and usually evenly coloured; dorsal line flexed forwards or straight at the apex; and peculiar host. **Similar species**: somewhat similar to *O. inulae*, but differs in having a longer inflorescence, and in the morphology of the bracts and the shape of the corolla (tubular to narrowly campanulate), dorsal line regularly arched-curved and lobes; and in particular a much higher filament insertion; as well as in host species. *Orobanche javakhetica* may co-occur but differs in its much shorter calyx, corolla, bracts, colour and host.

Orobanche arpica
Armenia (left, right)

NAME ***Orobanche artemisiae-campestris*** Vaucher ex Gaudin
SYNONYM ***Orobanche artemisiae*** Gren.

Habit: slender or slightly robust; inflorescence ± lax. **Height**: 10–30 cm **Stem**: simple, ranging yellowish-cream to mauve; with abundant glandular hairs; scales 10–20 mm, lanceolate, yellowish-cream or darker, with glandular hairs. **Calyx**: 10–15 mm, segments entire or unevenly divided; acuminate; yellowish-cream or pink, with glandular hairs; floral bracts equalling or slightly shorter than the corolla. **Corolla**: 16–20 mm, erect-spreading, ± tubular, curved with glandular hairs, yellowish to cream with reddish-violet veins; lobes rounded, slightly denticulate. **Carpel**: stigma dark mauve. **Stamens**: filaments inserted 3–4 mm above base of corolla, glabrous towards the apex, with long non-glandular hairs towards the base. **Host**: strictly parasitic on *Artemisia campestris* (Asteraceae). **Ecology and distribution**: distribution not fully known: from western, southern to central Europe (Spain, France, Germany, Czech Republic and Slovakia), in grassy pastures, on rocky slopes, and in xerothermic grassland. Flowering usually June–July. **Key diagnostics**: yellowish corolla with reddish colouration, open throat, calyx-teeth divided up ⅔, almost to the base and host-specific. **Similar species**: *Orobanche artemisiae-campestris* is frequently confused with *O. picridis*, which overlaps in distribution, but the former is easily distinguished by having fewer, larger, yellowish flowers, with reddish markings (rather than small, straight-back, pale corolla with contrasting dark stigmas and filiform calyx segments).

Orobanche artemisiae-campestris Czech Republic (above, right) photos: František Lamla

NAME ***Orobanche australis*** Moris ex Bertol.
SYNONYM ***Orobanche thapsoides*** Lojac.

Habit: inflorescence congested on all or most of its axis. **Height**: 20–40 cm. **Stem**: usually very robust; scales triangular. **Calyx**: lower segments well-developed; segments triangular. Floral bracts markedly longer than the corolla. **Corolla**: dorsal line rather straight, then geniculate at the tip; ventral line sharply bent at the lower lip; tube with short, straight glandular hairs; corolla <0.20 cm long, whitish-cream with fine violet veins; upper lip exceeding the lower lip and with lobes usually directed forward; all lobes crenate. **Carpel**: stigma yellow. **Stamens**: inserted ca. 0.2 cm above base of the corolla tube; basally with numerous long hairs; glabrous above. **Host**: herbs in the Asteraceae. **Ecology and distribution**: Sardinia, Sicily, peninsular Italy and southern France, in pastures from sea level to about 1,000 m. **Key diagnostics**: belonging to the complex Subsection *Minores* and similar to *O. canescens*, *O. pubescens*, *O. palaestina* and *O. amethystea*. *Orobanche australis* forms a dense, robust spike of flowers along its entire length (rather than laxer below); *O. australis* can be differentiated from *O. canescens*, *O. pubescens* and *O. amethystea* by the considerably long floral bract, a trait it shares with *O. palaestina* Reut. (not covered here). It can be distinguished from *O. palaestina* by its straight glandular hairs on the corolla (*O. palaestina* has a glabrous or sparsely hairy corolla).

Orobanche australis
Sicily, Italy (right)

NAME ***Orobanche austrohispanica*** M.J.Y.Foley
SYNONYM ***Orobanche gracilis*** subsp. ***deludens*** (Beck) A.Pujadas

Habit: robust; inflorescence ± lax, congested in the upper part. **Height**: 20–30 cm. **Stem**: simple, 0.5–1.0 cm across below the inflorescence, reddish-yellow, glandular-pubescent; scales 1.5–2.0 cm long, ca. 0.5–0.8 cm wide at the base, noticeably wide in proportion to length, very broadly lanceolate, pale yellow, glandular-pubescent. **Calyx**: lateral segments equally bifid, tapering, pink-yellow, glandular-pubescent; 0.9–1.5 cm long. **Corolla**: erecto-patent, ± campanulate, but ± keeled along the dorsal line, slightly galeate, distinctly curved proximally, less so distally; tube 1.8–2.5 cm long, cream to pale yellow or very pale brown and with longitudinal contrasting red veins on both inner and outer surfaces; upper lip slightly bilobed; lower lip lobes divergent, the central lobe rounded, laterals more acute, margins ciliate, lobes of both upper and lower lips denticulate. **Carpel**: stigma lobes ± touching, bright yellow; style pale pink with a few, scattered glandular hairs. **Stamens**: filaments inserted 1.5–2.5 mm above the base of the corolla, ± glabrous throughout their length and at the junction with the corolla base; anthers glabrous. **Host**: *Ulex* spp. (Fabaceae); perhaps other shrubby Fabaceae. **Ecology and distribution**: scrub, road verges, woodland clearings. Often on calcareous, but also on ultramafic, substrates 0–1,700 m. Restricted to southern Spain. Its presence in Morocco is likely based on a poorly preserved specimen collected at Beni Hosmar that may refer to this taxon. Flowers between April and May (possibly as early as March or as late as June). **Similar species**: very similar to *O. gracilis* (and considered a form thereof by some); differs in its ± dorsally keeled, distinctively and contrastingly reddish-veined, pale corollas, which lack the shiny red interior of *O. gracilis* and in its almost glabrous filaments and filament bases. Differs from *O. rapum-genistae* in its strongly veined corolla, ± glabrous filaments, lower filament insertion and less divergent stigma lobes. There are two other similar taxa with ± glabrous filaments. *Orobanche gracilis* var. *wagnerí* Beck was based upon a Hungarian specimen that differs in having a dark red-purple corolla and an appreciably glandular-pilose style. The other, *O. tetuanensis* Ball, was described from material collected at Tetuan, Morocco. This also has a dark red-purple corolla but also differs from *O. austrohispanica* in having a dense inflorescence and entire calyx segments.

Orobanche austrohispanica
Spain (left, right)
photos: Gonzalo Moreno Moral

NAME ***Orobanche ballotae*** A.Pujadas

Habit: long, lax inflorescence; spike (10–)13–19(–28) × 2–2.8 cm, cylindrical, ± rounded at the apex, multifloral, lax at least in its lower ½. **Height**: 20–40 cm. **Stem**: slender, whitish, sometimes purple-tinged, thickened at the base. **Calyx**: exceeded by the corolla tube, hairy-glandular, white, the apex sometimes slightly purplish but without conspicuous venation; segments free, contiguous or briefly imbricate (rarely fused at the base), subentire to very unequally bidentate; typically the lower segments very unequally bidentate and the upper subentire; 0.5–0.9 cm long, 0.25–0.35 cm wide. **Corolla**: erect to erecto-patent, tubular, the base slightly inflated; exterior tube glandular-pubescent, denser towards the limb and glabrous or glabrescent at the middle; internally hairy at the base; tube 1.0–1.4(–1.5) cm long, white (rarely with faint purple veins). Upper lip bilobed, scarcely emarginate, porrect; lower lip with median lobe larger than the laterals, irregularly toothed, non-ciliate. **Carpel**: stigma milk-white; style 0.6–0.7 cm with scattered glandular hairs or glabrous, white or sometimes faintly amethyst. **Stamens**: filaments covered with few glandular hairs, the lower ⅔ hairy; anthers 1.5–1.8 mm long, purple mottled. **Host**: *Ballota hirsuta* (Lamiaceae); a population on *Lamium garganicum* reported in Greece. **Ecology and distribution**: degraded xerophytic substrates in disturbed habitats between 20 and 1,500 m. Flowering April–May, bears fruit May–June. Southern Spain and probably north-west Africa; populations also reported in Pelopónnisos region, on Mt Mavro Oros, Greece. **Similar species**: distinguished from *O. minor* and relatives by the whitish colour of the whole plant, the segments of the calyx smaller and broader, corolla size small and strongly curved; anthers long, ellipsoid, always puberulous with a white suture base; the milky stigma is also key.

Orobanche ballotae Spain (above) photo: Enrique Triano, Spain (right)

NAME ***Orobanche balsensis*** (J.A.Guim.) Carlón, M.Laínz, Moreno Mor. & Ó.Sánchez

SYNONYMS ***Orobanche loricata* var. *synomora* subvar. *balsensis*** J.A.Guim.; ***Orobanche loricata* f. *balsensis*** (J.A.Guim.) Beck

Habit: slender, lax inflorescence, rather short relative to stem height. **Height**: (14–)17–34(–37.5) cm. **Stem**: purplish or creamy (somewhat purple-tinged towards the inflorescence), glandular-hairy throughout (hair density decreasing from the inflorescence towards the base); scales 1.0–1.5 × 0.23–0.5 cm, triangular-lanceolate. **Calyx**: segments free, ± unequal with upper segment longer, pink-purplish, densely covered with glandular hairs; (0.9–)1.1–1.25(–1.4) × (0.21–)0.25–0.35(–0.4) cm; all lobes slender to filiform; floral bracts just exceeding the corolla. **Corolla**: patent, ± narrowly tubular, slightly broadened distally, dorsal line geniculate at the point where the stamens are inserted and then straight; outer surface (especially in the dorsal area) densely covered with glandular hairs; (1.5–)1.6–2.0 cm long; whitish-cream with purple veins; upper lip shallowly bilobate, with lobes variably erect, forward-directed or porrect, ± rounded, with undulated and crenate-denticulate margins; lower lip lobes deflexed, almost overlapping or slightly distant, rounded to sub-quadrangular with ± crenate-denticulate margins. **Carpel**: stigma pinkish-purple; style white and glabrous at the base, increasingly purplish and covered with short glandular hairs towards the apex. **Stamens**: filaments (0.65–)0.85–0.9(–1.0) cm, inserted at (0.21–)0.25–0.35(–0.4) cm from the base of the corolla, hairy only in the basal ½, whitish or somewhat yellowish and ± purple-tinged below the anthers. Anthers 1.5–1.75 × 0.7–0.9 mm, purple or brown-greyish (beak white, conspicuous), glabrous except along the sutures in the basal ½. **Host**: *Carlina corymbosa* (Asteraceae), possibly other *Carlina* species. **Ecology and distribution**: widespread but rather uncommon: south-west Portugal and Spain; probably southern France, Corsica and Sicily, at elevations of 0–900 m; dunes, sea cliffs and maquis. **Key diagnostics**: pale plant; corolla with straight dorsal line and slender to filiform calyx lobes; specific to *Carlina*. **Similar species**: a plant defined more by ecology than morphology: very similar to *O. picridis* and difficult to distinguish from that species other than by host range; *O. canescens* is known to grow on similar hosts but this Sicilian taxon differs by its much shorter, erecto-patent and dorsally curved flowers; co-occurs with *O. minor* and *O. calendulae* in south-west Portugal; *O. minor* has stronger pigmentation and smoothly arched corolla tube; *O. calendulae* has a different corolla profile and host (*Calendula*).

Orobanche balsensis
Portugal (right, page 268, 269)

NAME ***Orobanche bartlingii*** Griseb.
SYNONYMS ***Orobanche alsatica* subsp. *libanotidis*** (Rupr.) Pusch; ***Orobanche libanotidis*** Rupr.

Habit: usually robust; inflorescence compact in the upper portion; stems naked below. **Height**: 15–40(–60) cm. **Stem**: stout, rarely slender, erect, glandular-pubescent, yellowish, yellowish-brown or pinkish; scales ovate to lanceolate-triangular. Inflorescence cylindrical to ovate, ± dense, usually shorter than its stem. **Calyx**: segments bidentate, short, usually subulate at tip, glandular-pubescent; 6–10 mm. Floral bracts lanceolate, about ⅔ of the length of the corolla or sometimes longer. **Corolla**: tubular, corolla dorsal line evenly curved at the proximal and distal part, almost straight at the middle; flowers held horizontally; 12–17(–18) mm, tube yellow, yellowish-brown, yellowish-pink, reddish-purple to reddish-brown; upper lip almost entire or slightly emarginate, with broad and erect lobes; lower lip 3-lobed, irregularly crenate and deflexed. **Carpel**: stigma bilobed with lobes just touching, yellow or rarely orange; style glabrous or rarely sparsely glandular-pubescent. **Stamens**: filaments inserted 1–3 mm above the corolla base, with long hairs at the basal ½ and sparsely glandular-pubescent under the anthers; anthers usually pubescent at the line of fusion. **Host**: mainly *Seseli* spp., such as *S. libanotis* and *S. transcaucasicum*, rarely *Bilacunaria microcarpos* (Apiaceae). **Ecology and distribution**: rocky, partially shaded slopes, scrub and forests, xerothermic grasslands, mountain meadows and steppe hills. Flowering (end May) June–July, fruiting July–August. A Eurasian species, occurring from the Pyrenees in Spain through France, Germany and central Europe, and more frequent in eastern Europe, the Baltic states (Estonia) and Russia (European and Asiatic), to Kazakhstan, Kyrgyzstan, Turkey and the Caucasus. **Key diagnostics**: usually tall and brightly coloured plant, in shades of yellow and pink, with a dense inflorescence, relatively short corolla 12–18 mm long, with dorsal line evenly curved at the proximal and distal part, almost straight at the middle, and characteristic host. **Similar species**: *Orobanche alsatica* is morphologically similar and closely related to *O. bartlingii*; the former has a longer corolla (20–25 mm) with regularly and strongly curved dorsal line, glandular-pubescent style, and stamens inserted 4–7 mm above the base of the corolla tube; it parasitises *Peucedanum* (rather than *Seseli*).

Orobanche bartlingii
Poland (right, page 272, 273)

NAME ***Orobanche baumanniorum*** **Greuter**

Habit: a short, stocky plant, sometimes clumped, with thick stems with numerous long scales below, and rather small flowers relative to the stem. **Height**: 10–25 cm. **Stem**: dull dark red to purplish, inflorescence extends over most of the stem; densely glandular-pubescent; lower scales brown, wavy, narrowly triangular. **Calyx**: normally 2-lobed, very long, tapering; segments filiform and exceeding the corolla tube; 10–16 mm. Floral bracts ± exceeding the corolla tube. **Corolla**: sub-erect, very abruptly curved along the dorsal line distally; upper lip emarginate; lower lip with subequal, acute narrowly triangular lobes, all toothed; tube purplish-burgundy with pale pinkish-cream on the interior, veined with darker pink; corolla 15–20 mm long. **Carpel**: stigma red to dark purple. **Host**: Caprifoliaceae: mostly *Pterocephalus pterocephala*, rarely *Scabiosa*, *Cephalaria* or *Lonicera*. **Ecology and distribution**: sunny, dry slopes in open woods, *Pinus* and *Abies* forests from south-east Italy, Greece and Albania to Turkey. **Key diagnostics**: small, stocky, dark red spikes; very abruptly curved corolla dark red externally, pale internally, with acute, very narrowly triangular lobes. **Similar species**: none; *O. sanguinea* has overlapping distribution and similar colouration, but very different corolla characteristics, host and ecology.

Orobanche baumanniorum Greece (above, right) photos: Toby Nowlan

NAME ***Orobanche boninsimae*** (Maxim.) Tuyama
SYNONYM ***Platypholis boninsimae*** Maxim.

Habit: small but robust, the whole aerial parts vivid yellow; inflorescence cylindrical subglobose, congested with 10–30 flowers. **Height**: 10–15 cm. **Stem**: base much-branched or simple, erect, often thickened below, with multicellular glandular hairs; scales sparse, oblong, glabrous. **Calyx**: divided into 2 lateral sepals, oval or oblong bidentate, scarcely ½ the length of the corolla, with subacute teeth, glandular-hairy. Floral bracts oval, glandular-hairy, equalling the length of the calyx, obtuse. **Corolla**: 2.0–2.5 cm long, tubular, inflated, bilabiate, bright yellow with stiffer petals, glandular-pubescent, the tube of the flower slightly arched at the back in the middle, slightly wrinkled. Upper lip obtuse, erect, often emarginate with a slightly curved margin, lower lip 3-lobed, lobes triangular, subequal, obtuse at the apex. **Carpel**: stigma 2-lobed, yellow; style glabrous. **Stamens**: filaments glabrous, inserted in the tube of the corolla, strongly exserted, sub-thick; anthers bilocular, thickly attached connective tissue, obtuse. **Host**: woody species, *Ochrosia* (Apocynaceae), *Melicope* (Rutaceae) and *Bischofia* (Phyllanthaceae); other hosts that need confirmation are *Callicarpa* (Lamiaceae) and *Ardisia* (Primulaceae). **Ecology and distribution**: endemic to the Bonin-Islands (Ogasawara Islands) about 1,000 km south-east of Japan, where it grows in shady, moist forests. Flowering February–March. Pollinated probably by passerine birds. **Key diagnostics**: small, bright yellow plant, anthers and stigma protruding from the corolla, exserted stamens (unique within the genus); stem strongly branched at the caudex. **Similar species**: *Phacellanthus tubiflorus* or other species with clusters of flowers at ground-level on the forest floor are somewhat similar but other morphological features of *O. boninsimae* stressed above are unique. Until recently, the plant was classified in its own genus, *Platypholis*. Phylogenetically it is closest to *O. coerulescens*.

Orobanche boninsimae Japan (above, right) photos: Akihiro Nishimura

NAME ***Orobanche calendulae*** Pomel

SYNONYMS ***Orobanche artemisiae-campestris* subsp. *calendulae*** (Pomel) O.Bolòs, Vigo, Masalles & Ninot; ***Orobanche mauretanica* subsp. *calendulae*** (Pomel) Cout.; ***Orobanche minor* var. *calendulae*** (Pomel) Batt. & Trab.

Habit: a short, slender to somewhat robust plant with flowers from ground-level, laxer below. **Stem**: slender to robust, brownish-red or violet; flowers ± lax, over most of the stem, 10–30(–50) cm; lower scales few, 10–15 mm, ± lanceolate, yellow-brown, with glandular hairs. **Calyx**: 10–15 mm, segments free, equal or unequal. Floral bracts 10–15 mm, slightly shorter than the corolla. **Corolla**: 15–18 mm, erect-spreading, tubular, markedly curved; margins somewhat denticulate, with glandular hairs, yellowish, with reddish or purplish veins. **Carpel**: stigma purple, sometimes orange-yellow. **Stamens**: filaments inserted 2–4 mm above corolla base; ± glabrous except at the base. **Host**: parasitic on *Calendula suffruticosa*; possibly other Asteraceae. **Ecology and distribution**: sea cliffs in Madeira, Portugal, south-west Spain and northern and north-west Africa; records from the Canary Islands on *Andryala* refer to *O. andryalae*. **Key diagnostics**: many-flowered stems, curved, pale yellow (or pink) corolla; host range and ecology. **Similar species**: a somewhat confused taxon belonging to the complex Subsection *Minores*, originally documented in Algeria. Our photograph depicts a plant on a sea cliff in Portugal growing on *C. suffruticosa* that may be the plant described by Beck as *O. mauretanica* var. *calendulae* (≡ *O. calendulae*). These plants possess pale orange stems, a whitish corolla with violet pigmentation, reddish stigma, apically filiform calyx lobes, and flowers that are arranged over most of the stem. These features appear to be broadly consistent with the holotype of *O. calendulae*, but the relationships among these coastal *Calendula*-specific plants are unclear. *Orobanche minor* typically has fewer, laxer flowers and a more narrowly tubular, evenly arched corolla.

Orobanche calendulae
Madeira (left)
photo: Miguel Sequeira
Portugal (right)

NAME ***Orobanche camptolepis*** Boiss. & Reut.
SYNONYMS ***Orobanche hermonis*** Mouterde; ***Orobanche sideana*** Gilli

Habit: short, robust, oblong or short cylindrical, fairly dense inflorescences with flowers from ground-level. **Height**: 8–15(–20) cm. **Stem**: pinkish, pale violet or yellowish, glandular-pubescent, the base strongly thickened, the tip often attenuated with short, lanceolate cusp, hooked when dry; scales numerous, up to 10 mm long, broadly ovate to suborbicular, thick, brownish. **Calyx**: segments free, fairly broad, bidentate up to ½ the length, with lanceolate-subulate segments up to 10 mm long, shorter than the corolla tube. Floral bracts up to 10 mm long, suborbicular or broadly ovate, somewhat clasping flowers. **Corolla**: tubular, sub-erect, the dorsal line usually almost straight from a somewhat curved base, glandular-pilose; 12–18 mm; variable in colour – pale violet, pinkish or light blue, violet, yellowish, or yellowish-brown, pinkish when dry; upper lip slightly 2-lobed, lobes broad, shallow, short and deflexed; lower lip 3-lobed, lobes usually mucronate, acute, middle larger, all abruptly pointed. **Carpel**: stigma pink or whitish, style sub-glabrous. **Stamens**: filaments inserted 3–5 mm above corolla tube; glabrous or with few hairs at base; anthers sub-glabrous. **Host**: *Polygonum* spp. (Polygonaceae). **Ecology and distribution**: rocky, sandy and loamy slopes, semi-deserts, steppes. Flowering June–August. The range requires further research; known from Turkey in Europe, western Asia (Afghanistan, Iran, Iraq, Israel and Palestine, Lebanon, Syria and Turkey) and central Asia (Kyrgyzstan, Tajikistan, Turkmenistan and Uzbekistan) to the Arabian Peninsula. **Key diagnostics**: variable, but easily recognised by the stature, numerous broadly ovate scales with subulate, curved tips; floral bracts suborbicular or broadly ovate with subulate, curved tip, somewhat clasping flowers; inflorescence very dense; tubular corolla almost enclosed in the bract, ± straight, sub-erect, not geniculate (as in *O. cernua*), and the usually abruptly mucronate lobes of the lower lip. **Similar species**: none; *O. sideana* and *O. hermonis* are considered by some authors to be yellow and highland forms of the species, respectively.

Orobanche camptolepis Israel (above) photo: Yael Orgad, (right) photo: Avinoam Danin

NAME ***Orobanche canescens*** C.Presl
SYNONYMS ***Orobanche flavidiflora*** Guss.; ***Orobanche lucana*** N.Terracc.

Habit: slender, with flowers rather lax, but congested above. **Height**: 10–30(–50) cm tall. **Stem**: angulate, glandular-puberulent, pinkish or yellowish; scales lax and ovate-lanceolate. **Calyx**: unequally bifid, with segments linear-acuminate, highly glandular-pubescent, 0.7–1.2 cm long. Floral bracts pubescent, 1.2–2.0 cm long, lanceolate, obtuse, equalling or exceeding the corolla. **Corolla**: dorsal line evenly curved over entire length, or slightly downward-bent at the apex, 1.0–1.9 cm, creamy yellow, often with purple veins near the apex; upper lip emarginate with lobes directed forward; lower lip lobes directed forward, subequal or the central slightly longer than the laterals; all lobes rounded and denticulate. **Carpel**: stigma yellow or whitish; style glabrous. **Stamens**: inserted 2–3 mm from corolla base, filaments densely hairy below. **Host**: *Galactites tomentosus*, *Carlina nebrodensis*, *C. sicula*, *Glebionis coronaria* and various other Asteraceae. **Ecology and distribution**: coastal plains and hillslopes, occasionally inland, northern Africa (Algeria and Tunisia), Italy (especially Sicily), reported from Greece (incl. Crete – possibly in error for *O. pubescens*), Corsica and Sardinia. **Key diagnostics**: densely hairy filament bases and pale stigma; yellowish corolla with reddish tinge. **Similar species**: belonging to the complex Subsection *Minores* and often misidentified. It can be distinguished from *O. minor*, *O. pubescens* and *O. litorea* by the densely hairy filament bases, the acute angle of insertion of the corolla and the yellowish or whitish stigma. It is differentiated from *O. hederae*, *O. artemisiae-campestris* and *O. amethystea* by the characters stressed above and the corolla profile. *Orobanche esulae*, a parasite of *Euphorbia esula* from the Balkans (Bulgaria and Serbia), is similar in colour but differs in its very high stamen insertion, 4–6 mm, and host.

Orobanche canescens
Sardinia, Italy (left, right, page 284, 285)

NAME ***Orobanche caryophyllacea*** Sm.
SYNONYMS ***Orobanche laxiflora*** Rchb.; ***Orobanche galii*** Duby; ***Orobanche vulgaris*** Poir.

Habit: rather robust, curved stems, few-flowered, lax. **Height**: 15–40(–50) cm. **Stem**: glandular-hairy, whitish, pale yellow or pale pink in colour with dark brown scales at the very base. **Calyx**: entire to ± equally bifid, 1.0–1.7 cm. Floral bracts ovate-lanceolate, acuminate, shorter than the flowers. **Corolla**: densely glandular-hairy, erecto-patent, broadly campanulate, broadest just behind mouth, ± uniformly curved; rather large, 2.0–3.2 cm, whitish, pale yellow or pale pink, to bright brownish-violet; upper lip notched with the lobes erect upon opening; lower lip with subequal glandular-ciliate almost fringed lobes. **Carpel**: stigma dark purple, rarely orange or yellow, lobes distant. **Stamens**: filaments inserted 1–3 mm above base of the corolla tube, hairy below, glandular above. **Host**: Rubiaceae, primarily *Galium* species. **Ecology and distribution**: widespread from the mountains of North Africa, through montane areas in the Mediterranean region, northwards to southern Norway, and eastwards through central Europe, Asia Minor, the Caucasus to Iran and central Asia. In central Europe, found in semi-arid grasslands and meadows, roadsides and open woodland, on loose calcareous loam and loess soils. In the British Isles, it occurs on undercliffs and fixed dunes. In western Asia, in rocky, loamy, grassy slopes, steppe, edges of forests, pine groves. Flowering (end of April) May–June. **Key diagnostics**: pale, large-flowered plant with a broadly campanulate corolla and dark, separated stigma lobes; flowers smell strongly of cloves. **Similar species**: most similar to *O. clausonis*, which is also a parasite of Rubiaceae from the western Mediterranean but differs in having a more narrowly tubular corolla and a shorter calyx and corolla; *O. teucrii* is a similar species that grows on *Teucrium* and is widespread in central Europe – it lacks the fragrant flowers; *O. lutea* is also a similar continental species that differs mainly in having yellow stigma lobes.

Orobanche caryophyllacea
Poland (left, page 288)
Holland (right, page 289)
photos: Ted Smit

NAME ***Orobanche centaurina*** Bertol.
SYNONYMS ***Orobanche echinopis*** Pančić; ***Orobanche kochii*** F.W.Schultz; ***Orobanche ritro*** Gren.

Habit: robust, many-flowered sub-cylindrical spikes. **Height**: (20–)30–40(–60) cm. **Stem**: stout, bulb- to clavate-shaped at the base, reddish, pink or orange (all yellow in ***f. citrina***), glandular-pubescent; scales wide in proportion to length, upper scales lanceolate-ovate to ovate-triangular, 15–22 mm long, the lowermost ± triangular, 8–16 mm long. **Calyx**: segments free or occasionally to 2 mm connate at base, asymmetrically ovate and unequally bidentate, 12–16 mm long. Floral bracts as long as the corolla or shorter. **Corolla**: sub-tubular, the dorsal line unevenly curved (curved at base, then almost straight), (15–)20–28 mm, pale pink or yellow to carrot-red, glandular-pubescent; upper lip entire to emarginate, patent, wide; lower lip 3-lobed, lobes are big, divergent, spathulate, concave. **Carpel**: stigma lobes touching, yellow; style sparsely glandular-pubescent. **Stamens**: filaments inserted 4–6 mm above the corolla base, hairy in the lower ½, sparsely glandular above; anthers glabrous or very shortly hairy along the sutures. **Host**: mainly on *Centaurea* (especially *C. scabiosa*), occasionally *Echinops*, *Ptilostemon* and *Jurinea* (Asteraceae). **Ecology and distribution**: xerothermic grasslands, steppe, thermophilous scrub, forest and field margins and ruderal sites. Flowering period (late June) July–August. A Eurasian, continental species, from France, central and southern Europe to central Asia, central China and India. Recently split from *O. elatior*; distribution in the east of the range requires verification. **Key diagnostics**: distinct carrot-red to whitish-rosaceous colouration, and large, divergent, spathulate, concave lower-lip lobes with a patent, wide upper-lip. **Similar species**: formerly, *O. centaurina* (also known as *O. kochii*) was included in *O. elatior* (which also grows on *Centaurea scabiosa*). *Orobanche elatior* s.str. is typically large, yellowish, ochre to light brown, with dense flowers, the corolla regularly curved throughout, with long and linear-lanceolate bracts. Close to *O. staehelinae*, as well as *O. loscosii*, which has regularly curved flowers without a straight dorsal line, darker (pinkish) stigma and a lower filament insertion.

Orobanchce centaurina
Czech Republic (right) photo: František Lamla
Poland (page 292)
Orobanchce centaurina f. *citrina*
Poland (page 293)

NAME **Orobanche cernua** Loefl.
SYNONYMS **Orobanche berthelotii** Webb & Berthel.; **Orobanche hispanica** Boiss.

Habit: short, slender to robust, congested. **Height**: 15–35 cm. **Stem**: usually thick, with a diameter of (0.3–)0.5–0.8(–0.9) cm, bulbous at the base (0.9–2.0 cm), glandular-pubescent, yellowish to pale violet; scales deltate below (0.8–1.1 × 0.5–0.8 cm), and ovate-lanceolate above (0.6–1.1 × 0.4–0.7 cm). **Calyx**: segments free, bifid (cleft 0.25–0.4 cm) to deeply bifid (cleft up to 0.7 cm), sometimes bidentate (cleft up to 0.25 cm), seldom entire, with subulate segments 0.6–1.0 cm long. Floral bracts (0.5–)0.7–0.9(–1.2) × 0.35–0.5 cm, ovate to lanceolate, much shorter than the corolla tube. **Corolla**: inflated at the base, constricted and straight in the middle, strongly arched forward (cernuous), the dorsal line geniculated above the insertion of the stamens, sparsely glandular-pubescent; tube (1.3–)1.5–1.8 cm long, white at the base, blue, dark violet or purple distally; upper lip distinctly bilobed; lower lip 0.2–0.22 × 0.23–0.25 cm, subequal, obovate, not ciliate. **Carpel**: lobes touching, pale yellow or white; style with some glandular hairs, sometimes sub-glabrous or glabrous. **Stamens**: filaments 0.6–0.7 cm long, inserted obliquely just above the base of the corolla, glabrous or pubescent in the lower part. **Host**: parasitic usually on *Artemisia* spp., rarely on various herbs in the Asteraceae, Fabaceae, Amaranthaceae, Nitrariaceae and Solanaceae. **Ecology and distribution**: arid areas, sand dunes, xerothermic scrub, from sea level to 1,500 m. Flowering (end April) May–June. Very widespread from southern Europe to Turkey, northern Africa and the Canary Islands, Ethiopia, through the Arabian Peninsula, and into much of western Asia and eastern China; strongly southern. An endemic variety, var. *australiana* occurs in Australia. **Key diagnostics**: strongly cernuous corolla. **Similar species**: *Orobanche cumana* is similar but with longer and laxer inflorescences, paler (whitish) corolla and differing ecology. Plants from Australia, parasitic on native Asteraceae, with larger flowers (to 20 mm), have previously been recognised as *O. australiana* but are currently treated at the varietal level. As the only native *Orobanche* species in Australasia, a firmer grasp of its relationships and origins would benefit from further investigation.

Orobanche cernua
Israel (left, right, page 296)
In cultivation at Oxford Botanic Garden, UK (page 297) source: Lanzarote, Canary Islands

NAME ***Orobanche chironii*** Lojac.
SYNONYM ***Orobanche alsatica* var. *chironii*** (Lojac.) Bég.

Habit: slender, lax spikes. **Stem**: to 45 cm, distinctly swollen at base, most exterior parts slightly glandular-pubescent; scales to 10 mm, triangular-lanceolate. **Calyx**: segments free with lobes bidentate; 15 mm; purple, contrasting the pale corolla tube. Floral bracts 15–25 mm, lanceolate. **Corolla**: erecto-patent to patent on stem, shape narrowly campanulate and uniformly curved; 18–25 mm, cream or pale yellow, often tinged violet-pink; upper lip margin entire or weakly bilobed; lower lip with ciliate margin. **Carpel**: stigma purplish. **Stamens**: filaments hairy, inserted 3–4 mm above base of corolla; anthers glabrous. **Host**: *Anthemis cupaniana* (Asteraceae); also reported on *Opopanax chironium* (Apiaceae), possibly in error. **Ecology and distribution**: endemic to Sicily; cliffs and crags. **Key diagnostics**: narrowly campanulate corolla with glandular-ciliate lower lip, stem highly swollen at base, muted corolla colouration, filaments inserted more than 2 mm above corolla base. **Similar species**: belonging to the Subsection *Minores*, differing from similar taxa in the diagnostics emphasised above.

Orobanche chironii
Sicily, Italy (right)

NAME ***Orobanche cicerbitae*** (Uhlich & Rätzel) Tzvelev
SYNONYM ***Orobanche flava*** **subsp.** ***cicerbitae*** Uhlich & Rätzel

Habit: slender or slightly robust; inflorescences ± lax, much shorter than the stem beneath. **Height**: to 35(–50) cm. **Stem**: slender or slightly stout, pale yellowish-brown, yellow, pinkish-brown, glandular-pubescent; scales (9–)12–14 mm, broadly ovate-lanceolate, lax, glandular-pubescent. **Calyx**: segments free, entire, basally ovate; teeth narrowly lanceolate to filiform, 8–8.5 mm. Floral bracts up to 17 mm, shorter than or as long as the corolla tube. **Corolla**: broadly tubular, the dorsal line strongly curved (but ± deflexed distally), (21–)22–23(–25) × (6–)7(–8) mm, whitish, yellowish, yellow, slightly rose-coloured or pale brownish; upper lip broadly 2-lobed to emarginate, lobes porrect at first, later patent; lower lip with subequal, large and spathulate lobes with rounded and ± finely irregularly denticulate and glandular margins. **Carpel**: stigma with broad, touching subspherical lobes; yellow; style with scattered glandular hairs. **Stamens**: filaments inserted 3 mm above the corolla base; basally and upwards numerous long hairy, below anthers glandular-hairy; anthers hairy basally and along the suture. **Host**: *Caucasalia macrophylla*, *C. pontica*, *Pojarkovia pojarkovae*, *Prenanthes petiolata* and *Senecio propinquus* (Asteraceae). **Ecology and distribution**: humid stands of tall herbaceous vegetation in subalpine meadows, forests and scrub or forest margins, near river valleys, waterfalls and rocky slopes, 1,600–2,300(–2,500) m. Flowering (June) July (August), fruiting end July–August. A Caucasian endemic: Russia, Azerbaijan, Georgia. **Key diagnostics**: pale yellowish, pinkish or whitish, very sensitive to damage; inflorescence shorter than the remaining part of the stem; corolla ± tubular, with upper lip ± porrect and strongly constricted below the filament insertion point; stigma bright yellow. **Similar species**: some authors include this species under *O. flava*, which is taller, more pigmented, with longer and denser stem scales, and with shorter corolla with dorsal line evenly curved, has different lobe morphology, glabrous style and a higher filament insertion, as well as a different host range. *Orobanche mlokosiewiczii* differs in having a campanulate and slightly curved corolla, as well as in lobe morphology and host range.

Orobanche cicerbitae
Georgia (left, right)

NAME ***Orobanche clausonis*** **Pomel**
SYNONYM ***Orobanche galii* var. *clausonis*** **(Pomel) Batt.**

Habit: slightly robust or slender; inflorescence 4–14(–20) cm, cylindrical, rounded at the apex, rarely sub-comose, fairly lax to dense. **Height**: 14–50 cm. **Stem**: scales sparse, 12–17 mm, lanceolate to linear-lanceolate, glandular-pubescent, pale yellow, creamy. **Calyx**: segments divided approximately ½ their length into 2 broad, short, equal teeth 7–9 mm long, with glanduliferous hairs. Floral bracts 11–20 mm, lanceolate, glandular-pubescent. **Corolla**: (14–)15–19(–22) mm, erecto-patent, ± tubular to slightly campanulate, dorsal line curved, somewhat swollen at the base, generally straight back and galeate towards the limb; light yellow, creamy, with glandular hairs. Upper lip slightly bilobed or entire, short, lobes straight sometimes patent; lower lip obtuse, rounded to oblong, subequal, recurved, somewhat serrated or crenate. **Carpel**: stigma bilobed, red or pink; style glabrous, sometimes with sparse hairs below the stigma. **Stamens**: filaments inserted 2–4.5 mm above corolla tube base; sparsely hairy below, glabrous towards the base; anthers glabrous or with hairs at the suture. **Host**: exclusively on Rubiaceae, mainly *Rubia*, *Asperula* and *Galium*. **Ecology and distribution**: oak forests, thickets, meadows, roadsides, rocky slopes, usually at 400–1,600 m. Mediterranean in distribution (Algeria, Balearic Islands, Sicily, Malta, Morocco, Portugal, Spain and Tunisia). Flowering April–May (June). **Key diagnostics**: a pale plant, with a pale yellow, tubular to slightly campanulate corolla with a galeate lip and contrasting red stigma; host-specific. **Similar species**: closely related to the more widespread *O. caryophyllacea*, also a parasite of Rubiaceae, which differs in having a more broadly, long-campanulate corolla and a longer calyx.

Orobanche clausonis
Spain (left, right)
photos: Gonzalo Moreno Moral

NAME ***Orobanche coerulescens*** Stephan
SYNONYM ***Orobanche ammophila*** C.A.Mey.

Habit: short, robust. **Height**: 10–30(–40) cm. **Stem**: usually robust and thick, erect, yellowish, woolly; scales 8–15 mm, ovate-lanceolate to lanceolate, usually glabrous. Inflorescence cylindrical, dense, ± densely white arachnoid-villous. **Calyx**: segments entire or divided to ½, with teeth narrowly lanceolate and ± densely white arachnoid-villous, (8–)15–17 mm. Floral bracts usually shorter than the corolla, ovate, woolly. Flowers erecto-patent at first, bent forward or horizontal later. **Corolla**: tubular, inflated below insertion of the stamens, strongly curved forward from the base; (10–)15–20(–23) mm; blue, blue-whitish, blue-violet, rarely whitish or yellowish; corolla arachnoid-villous; upper lip 2-lobed, porrect, lobes are broad, short, ovate with slightly crenate margins, glabrous; lower lip 3-lobed, lobes crenate, ovate, middle lobe slightly longer, with white folds, not ciliate. **Carpel**: stigma whitish or rarely whitish-yellowish, whitish-bluish; style sparsely pubescent or glabrous. **Stamens**: filaments inserted 6–7 mm above base of the corolla tube, sparsely pubescent and glabrous above; anthers usually glabrous. **Host**: *Artemisia* spp., especially *A. campestris*; rarely *A. lercheana* or *A. japonica* (in Japan) (Asteraceae). **Ecology and distribution**: arid and semi-arid sandy grasslands, steppe, ruderal places, desert areas, river and coastal beaches, dry pinewoods, fallow land and cultivated fields. Flowering (end May) June–July, fruiting (June) July–August. A very widespread Eurasian species, occurring from central Europe (eastern Germany, Austria, and Poland), through eastern Europe, the Caucasus and central Asia to Korea and Japan. The centre of its range is associated with the steppe or sandy areas of central and eastern Asia. **Key diagnostics**: easily distinguished by the thick stem, inflorescence (normally) strongly white arachnoid-villous, numerous, blue and geniculate flowers, without bracteoles, and whitish stigma. **Similar species**: superficially similar to *Phelipanche* but without bracteoles; white or white-yellow flowered plants have been recognised as *O. korshinskyi* Novopokr., which is possibly a colour variant of *O. coerulescens*. Also similar to the poorly known *O. pycnostachya* Hance, which also parasitises *Artemisia* spp., and has pale yellow, blue-tinged flowers and occurs from south-eastern Siberia to China, Japan and North Korea.

Infraspecific taxa (selected)

Forma *nipponica* is similar in all parts but completely glabrous.

Orobanche coerulescens
in cultivation, UK (left)
Japan (right)
Poland (page 307)
Orobanche coerulescens f. *nipponica*
Japan (page 306)

NAME ***Orobanche colorata*** K.Koch
SYNONYM ***Orobanche anatolica*** Boiss. & Reut.

Habit: robust with dense, many-flowered, cylindrical inflorescence, usually longer than remaining stem. **Height**: 20–30(–40) cm. **Stem**: thick, orange to dark orangish-red, usually glabrous; scales dense, numerous, ovate to lanceolate, up to 15 mm long; inflorescence sub-glabrous to lanate to strongly villous. **Calyx**: segments bidentate, with 2 lanceolate teeth, up to 15 mm long; floral bracts equalling or longer than flowers, ovate to broadly lanceolate, glabrous or sparsely puberulent to villous. **Corolla**: tubular-campanulate, dorsal line concave, 16–22 mm, orange to dark orangish-red (sometimes pale orange or cream) and ± concolourous outside and in, with ± dark red calluses on the folds; corolla glabrous or sub-glabrous, sometimes lanate. Upper lip usually undivided, curved and helmet-shaped, limb long ciliate-pilose or lanate; lower lip lobes short, rounded, almost equal in size. **Carpel**: stigma lobes somewhat distant, yellow; style sub-glabrous. **Stamens**: filaments inserted 3–4 mm above corolla base, sub-glabrous; anthers glabrous. **Host**: *Salvia* spp. (Lamiaceae), especially *S. verticillata* and *S. nemorosa*, rarely *S. glutinosa*. **Ecology and distribution**: rocky, sandy, loamy or steppe slopes, roadsides, subalpine meadows, scrub and forest margins, rarely in anthropogenic habitats, usually at (500–)700–1,900(–2,200) m. Flowering May–June (July), fruiting July–August. Caucasus (Armenia, Georgia, Azerbaijan and Russia) and surrounding areas of Turkey, Iran and Iraq. **Key diagnostics**: strongly orange-red pigmented dense, many-flowered inflorescences; corolla dorsal line of the corolla concave, with ± dark red callus on folds, with subentire to galeate upper lip; host specificity for *Salvia*. **Similar species**: some authors consider *O. anatolica* to be distinct; the plant has whitish-woolly pubescence but is otherwise similar. *Orobanche turcica* G. Zare & Dönmez described from Turkey is similar somewhat to *O. anatolica* in habit, with a dense inflorescence, similar bract indumentum and host, but differs in having a gamosepalous calyx with 4 equal teeth, a pale pinkish-brown corolla with a straight dorsal line, corolla lobes lanate on the inner side and a higher filament insertion.

Orobanche colorata
Turkey (left) photo: Sibel Bostancıoğlu
Georgia (right, page 310, 311)

NAME ***Orobanche crenata*** Forssk.
SYNONYMS ***Orobanche pruinosa*** Lapeyr.; ***Orobanche speciosa*** DC.; ***Orobanche tommasinii*** Rchb.f.

Habit: tall, many-flowered inflorescences, sometimes in large numbers. **Height**: 15–80(–120) cm. **Stem**: slender, 0.5–2.5 cm across, pale yellow to dark purple, erect, swollen at the base; scales 15–25 mm, rather woolly, ovate-lanceolate to linear-lanceolate, acuminate. **Calyx**: 10–20 mm, unequally bifid. **Corolla**: campanulate, erecto-patent to horizontal; lips divergent, upper strongly bilobed forming a ± erect standard, the lower lip with large suborbicular lobes; (12–)20–30 mm, white with purplish veins especially on the lips, yellow at the base. Flowers fragrant (carnation-scented). **Carpel**: stigma lobes pale, either whitish, yellow-orange or pink (rarely purple). **Stamens**: filaments inserted 2–4 mm above corolla base, hairy at least below. **Host**: herbs in the Fabaceae, especially leguminous crop plants such as *Vicia faba*, rarely on Apiaceae and other families. **Ecology and distribution**: Mediterranean Basin, Middle East and North Africa. Flowers May–August. **Key diagnostics**: whitish, campanulate flowers with a conspicuous carnation scent, often on long, contrasting purple stems. **Similar species**: smaller specimens in natural (non-agricultural) habitats can be superficially similar to robust specimens of *O. minor* and *O. picridis*. When fresh, the strong scent immediately differentiates this species, as does the pale stigma and divergent corolla lips. When dry, the divergent corolla lobes and hairy (villous) stems and bracts are clear diagnostics.

Infraspecific taxa (selected)

Var. *callieri* was recently raised to species rank as *Orobanche callieri* (Tzvelev) Tzvelev. It differs from the typical species in having a thicker stem with a very dense and many-flowered inflorescence and grows on *Salvia aethiopis*. Known only from the type locality in Crimea.

Orobanche crenata
Malta (left, right, page 314)
photos: Stephen Mifsud
Spain (page 315)

NAME ***Orobanche cumana*** Wallr.

SYNONYMS ***Orobanche bicolor*** C.A.Mey.; ***Orobanche cernua* subsp. *cumana*** (Wallr.) Soó; ***Orobanche cernua* var. *cumana*** (Wallr.) Beck; ***Orobanche cernua* f. *cumana*** (Wallr.) Beck

Habit: slender with cylindrical or oblong, usually lax many-flowered inflorescences longer than their stems. **Height**: (10–)40–65 cm. **Stem**: ± gracile, bluish-violet, yellowish (sometimes whole plant yellow), sparsely glandular-pubescent or glabrous; scales lax, ovate to ovate-lanceolate. Inflorescence (15–)22–30(–38) cm, cylindrical or oblong. **Calyx**: segments bifid or rarely entire, free, ovate to ovate-lanceolate, to ½ as long as the corolla, (5–)7–12 mm. Floral bracts about ⅓ to ½ or as long as the corolla, ovate to ovate-lanceolate, acute or subulate. **Corolla**: erect, then horizontal, later downward-pointing; tubular; dorsal line of the corolla geniculate in the lower part and ± uniformly curved in the rest, (16–)19–22 mm, pale blue, blue, violet, white-blue (rarely yellow), sparsely glandular-pubescent or glabrous. Upper lip with 2 erecto-patent, denticulate and glabrous lobes; lower lip with 3 short lobes, rounded or acute, of almost equal sizes, deflexed or porrect. **Carpel**: stigma white, rarely white-bluish, white-yellowish; style sub-glabrous. **Stamens**: filaments inserted (4–)5–7 mm above corolla base; sub-glabrous or with sessile glandular hairs; anthers glabrous or sparsely hairy at the line of fusion. **Host**: a weed of cultivated plants, mainly *Helianthus annuus* (Asteraceae), less often *Nicotiana tabacum* and *Lycopersicon esculentum* (Solanaceae); also on wild hosts including *Artemisia* and *Xanthium* (Asteraceae). **Ecology and distribution**: cultivated fields, ruderal and inter-segetal habitats, steppe, semi-deserts and halophytic habitats. Flowering (end of April) May–July (September), fruiting July–August (September). Central Asia to south-east Europe; rarely in central Europe (e.g. Hungary and Slovakia); recently reported from North Africa (Morocco and Tunisia), as well as from North America (USA, Yakima in Washington). **Key diagnostics**: often on cultivated plants; flowers pale blue with short lobes; dorsal line of the corolla geniculate in the lower part and ± uniformly curved in the rest; ovate to ovate-lanceolate bracts; inflorescence is usually lax and longer than the stem beneath. **Similar species**: the relationship between *O. cernua* and *O. cumana* is unclear as both are polymorphic. *Orobanche cumana* parasitises cultivated plants, but also, like *O. cernua*, wild species. Typically *O. cumana* is taller and slender with a long, lax inflorescence with paler and longer flowers; these characteristics may be influenced by the host.

Orobanche cumana
Georgia (right)

NAME ***Orobanche cypria*** Reut. ex Kotschy

Habit: inflorescence rather stocky, compact, crowded, much longer than its stem. **Height**: 10–22 cm. **Stem**: robust, bright red to pinkish-red with dense white glandular hairs; scales rather numerous, broad and conspicuous. **Calyx**: fused, bifid, with subequal, broad lobes, acuminate, shorter than or equalling the corolla tube; floral bracts very broad at the base, rather paler than the stems and flowers, or dark blackish-maroon abruptly acuminate and brown and papery at the tips; calyx and bract with rather long glandular hairs. **Corolla**: erect, narrowly tubular with a smoothly curved dorsal line; small, 10–12 mm, ca. 4 mm wide at the apex, pink to maroon-red with darker veins; upper lip 2-lobed; lower lip with subequal lobes, all lobes finely fimbriate-denticulate. **Carpel**: stigma lobes touching, pink or pale orange-pink. **Stamens**: filaments inserted 3 mm above the base of the corolla tube, sparsely hairy below. **Host**: possibly only on *Pterocephalus* spp. (Caprifoliaceae), but also reported on *Cistus* (Cistaceae), *Salvia* and *Scutellaria* (Lamiaceae). **Ecology and distribution**: an endemic of Cyprus, mainly in the central mountains, 150–1,600 m, on rocky slopes and scree, and in open pine forests and vineyards; common in the Limassol district. Records in Iran and Iraq require confirmation. **Key diagnostics**: rather short, squat, congested, strongly dark red-pigmented spikes of flowers; corolla lobes with conspicuously plicate and fimbriate-denticulate margins; distinct ecology and host range. **Similar species**: *Orobanche baumanniorum* is similar in morphology, ecology and host range, and occurs in Greece and Turkey; this apparently closely related species has a more abruptly curved (to geniculate) corolla dorsal line, and lacks the finely toothed corolla lobe margins of *O. cypria.*

Orobanche cypria
Cyprus (right)
photo: Charalambos S. Christodoulou
(*Flora of Cyprus*)

NAME ***Orobanche dagestanica*** Ó.Sánchez, Piwow., Fateryga, Svirin & Murtaz.

Habit: robust; inflorescence cylindrical to somewhat ovoid, usually shorter than its stem. **Height**: (30–)36–40(–45) cm. **Stem**: pink to pale brownish-pink, thick, densely glandular-pubescent in the entire aerial part; scales (15–)18–24(–33) mm, dense in the lower part, ovate-lanceolate, triangulate-lanceolate to lanceolate. **Calyx**: segments simple, free, separate, entire, narrowly lanceolate, sometimes almost filiform, with long acuminate teeth with a ± ovate base, densely glandular-hairy; (8–)10–15(–20) mm long, usually shorter than ½ of the corolla tube. Floral bracts (13–)16–20(–21) mm, shorter or rarely almost equalling the corolla, ovate-lanceolate to lanceolate, dark pink to pale yellowish-brown, with dense white glandular hairs. **Corolla**: sub-erect, tubular-infundibuliform, widening at the throat, usually strongly inflated above the insertion of the filaments, and strongly constricted below; the dorsal line ± evenly curved at the base, straight at the middle part and bent forward at the apex; (21–)25–29(–31) mm long, (8–)10–11(–12) mm in diameter in the central part, pale pinkish-brown with darker veins, externally densely glandular-pubescent with white hairs; upper lip slightly emarginate, with 2 rounded lobes, porrect to patent or little upward-curved later, sparsely to densely glandular-hairy also in the inner part; lower lip with 3 obovate or ± rectangular to ovate-triangular lobes, erose or irregularly crenulate-denticulate on margins and sparsely glandular-hairy also in the inner part. **Carpel**: stigma lobes distant, yellow; style with sparse, short glandular hairs. **Stamens**: filaments obliquely inserted at (2–)4–5(–6) mm above the corolla base; densely villous in the base to middle part, not glandular, upper part with dense short, glandular hairs; anthers basally and along ¾ of suture distinctly and numerously hairy. **Host**: parasitic on *Lophiolepis horrida* or sometimes *L. isophylla* (Asteraceae). **Ecology and distribution**: known only from shale screes with sparse alpine vegetation at an elevation of 2,400–2,765 m. Flowering period June–July, fruiting July–August. Probably endemic to the Greater Caucasus, so far known from a few localities in Russia (Dagestan). **Key diagnostics**: pink, conspicuously hairy in almost all aerial parts; calyx segments free, entire, shorter than corolla tube; corolla up to 30 mm, tubular-infundibuliform; host-specific. **Similar species**: somewhat similar to *O. inulae* and *O. cicerbitae* but easily distinguished by the indumentum, pinkish to brownish colouration, larger corolla, alpine habitat and host.

Orobanche dagestanica
Dagestan, Russia (right)
photo: Ramazan Murtazaliev

NAME ***Orobanche densiflora*** Salzm. ex Bertol.

Habit: very robust, congested spikes with numerous flowers; inflorescences much longer than their stems. **Height**: 30–50 cm. **Stem**: pale whitish-cream (rarely deep dull red); scales to 22(–30) mm long, lanceolate. **Calyx**: bifid; segments unequal, (7–)11–15 mm. Floral bracts dark brown, contrasting with the inflorescence, much-exceeding the corolla and overtopping the flowers in bud. **Corolla**: sub-erect to spreading, tubular, not inflated and with a gently curved dorsal line; 10–18 mm long; upper lip emarginate; lower lip lobes rounded, all with minutely toothed margins; corolla whitish-cream (rarely dark red). **Carpel**: stigma lobes divergent, yellow (rarely red, in all-red specimens). **Stamens**: filaments inserted 1–1.5 mm above the base of the corolla tube, sub-glabrous below and with broadened, almost deltate bases; anthers glabrous. **Host**: *Lotus creticus*, rarely *Ononis* (Fabaceae). **Ecology and distribution**: a western Mediterranean element: Algeria, Morocco, Spain, Tunisia (also Portugal, where it is possibly confused with *O. foetida*); coastal dunes. **Key diagnostics**: cream-coloured (rarely all dark red), very robust, congested spikes with numerous flowers with contrasting dark brown bracts. **Similar species**: rare, all-yellow forms of *O. foetida* can be similar; this species has foetid flowers (although not always), hairy filaments without the broad, deltate bases, a higher point of insertion and a different host range (typically *Ononis*); molecular work supports their distinction.

Orobanche densiflora
Spain (left, right, page 324, 325)
photos: Gonzalo Moreno Moral

NAME ***Orobanche dominae*** El Mokni

Habit: tall, dense, many-flowered spikes. **Height**: (20–)40–70(–100) cm. **Stem**: stout, bulbous to clavate-shaped at the base, reddish, pale or dark pink, pale violet, or rarely yellow, glandular-pubescent; scales yellow-brown, broad, elongated-triangular, 18–22 mm long. Inflorescence 20–40(–70) cm long, cylindrical, lax to dense. **Calyx**: segments free or occasionally to 2 mm connate at base, asymmetrically ovate and unequally bidentate; 14–18 mm long. Floral bracts broad ovate-lanceolate, shorter or rarely equal to the corolla. **Corolla**: tubular to narrowly campanulate, inflated above the insertion of the filaments, the dorsal line of the corolla unevenly curved (dorsal line curved at base, then almost straight); 23–27 mm; yellowish-white, tinged violet, with dark violet veins, glandular-pubescent. Upper lip slightly emarginate with 2 broad lobes, porrect at first, later slightly patent; lower lip 3-lobed, lobes deflexed, subequal, or the central lobe slightly larger than the laterals with crenate to irregularly dentate and glandular margins. **Carpel**: stigma lobes touching, yellowish-orange to purplish; style glabrous or sparsely glandular-pubescent near the base. **Stamens**: filaments inserted 4–4.5 mm above the corolla base, sparsely glandular in the lower ½, glabrous above; anthers glabrous or very shortly hairy along the sutures. **Host**: mainly on *Onopordum espinae*, occasionally on *Cynara cardunculus* subsp. *cardunculus* (Asteraceae). **Ecology and distribution**: xerothermic grasslands, steppe, thermophilous scrub, fallow land, ruderal sites and olive groves. Flowering period end March–April. Known only from Tunisia. **Key diagnostics**: long, dense spikes and host specificity for *Onopordum*. **Similar species**: *Orobanche dominae* resembles other species with a southern Mediterranean range in terms of the robustness of its scape and the density of its spike: *O. mauretanica* and *O. australis*. It differs from these in its wider corolla, which is inflated above the filament insertion point.

Orobanche dominae
Tunisia (left, right)
photos: Ridha El Mokni

NAME ***Orobanche ebuli*** Huter & Rigo

Habit: somewhat robust, often clump-forming; inflorescence elongated with lax flowers when mature, with flowers occupying the upper ⅔ of the stem. **Height**: 25–70 cm (rarely 1 m or more). **Stem**: thickened at the base, yellow-purple, densely glandular-hairy; scales ovate-lanceolate, 13–20(–22) × 3–4 mm, similar to bracts. **Calyx**: segments 10–13 × 2–3.5 mm, entire rarely bifid, densely glandular-hairy mixed with short simple hairs along the margins. Floral bracts ovate-lanceolate, 18–20 × 2.5–3 mm, wide at the base, gradually narrowing towards the apex, with short and dense glandular hairs, light ochre at the base, brownish in the upper ½, often exceeding the lower lip of the corolla. Bracteoles present (an unusual feature): acuminate, 3–6 × 1 mm, with translucent and irregularly fringed margin, with abundant glandular hairs and few simple hairs, longer towards the base. Bracteole clearly observable in living plants, more difficult to perceive in herbarium specimens. **Corolla**: 21–23(–24) mm long, with an arched dorsal profile with dense yellowish and short glandular hairs, yellow-whitish at the base, ochre-purple or dark purplish-violet in the rest (rarely entirely purplish), unscented (rarely with a faint cocoa-like scent). Upper lip slightly retuse, crenulate, with cilia and short glands at the margins; lower lip 3-lobed, with crenulated, glandular-ciliated lobes, the central longer than the laterals. **Carpel**: stigma purple, very rarely yellow; style with scattered glandular hairs, more abundant at the top. **Stamens**: filaments inserted 2.2–3 mm from the base of the corolla, with dense simple hairs at the base and scattered glandular hairs in the rest, more numerous below the anthers; anthers glabrous. **Host**: exclusively on *Sambucus ebulus* (Viburnaceae). **Ecology and distribution**: an Italian endemic, exclusive to Lazio and Abruzzo; beech wood margins at 1,200–1,300(–1,600) m. Flowering June–end August. **Key diagnostics**: additional calycine bracteole; elongated, long and lax inflorescence with small flowers; purple corolla lobes contrasting with the pale yellow stem; usually occurring in extensive clumps; ecologically distinct. **Similar species**: unlikely to be confused with any other species; phylogenetically closely related to *O. haenseleri*, *O. lucorum*, *O. salviae* and *O. mayeri*.

Orobanche ebuli
Italy (left, right, page 331)
photos: Francesco Santoro
Italy (page 330)
photo: Mirco Gruppi

NAME ***Orobanche elatior*** Sutton
SYNONYM ***Orobanche centaureae-scabiosae*** F.W.Schultz

Habit: robust, tall spikes with many flowers; inflorescence cylindrical (up 30 cm), dense, usually equalling its stem. **Height**: (20–)40–60(–95) cm. **Stem**: stout, yellowish-brown, yellow, orange, brown or pinkish-brown, glandular-pubescent; scales 15–22 mm, long, numerous and dense, narrow, elongated, linear-lanceolate, glandular-pubescent. **Calyx**: segments with 2 unequal or subequal bidentate halves, highly connate at the base, about ⅓ to ½ the length of the corolla, glandular-pubescent; 10–14 mm. Floral bracts 15–25 mm, shorter or slightly longer than the flowers, lanceolate, sparsely glandular-pubescent or glabrous. **Corolla**: sub-erect to erecto-patent, broadly tubular to narrowly campanulate, the dorsal line strongly and regularly curved throughout; (18–)20–22(–25) mm; pale yellow to pale brown, rarely yellow, orange-brown, pale-rosaceous; glandular-pubescent. Upper lip usually undivided, entire or slightly emarginated, with 2 rounded, erect lobes; lower lip 3-lobed, lobes are elliptic, crenate, deflexed, of almost equal size. **Carpel**: stigma bilobed, yellow, orange; style sparsely glandular or glabrous. **Stamens**: filaments inserted 3–6 mm above the base of the corolla tube, pubescent at the base and sparsely glandular-pubescent below the anthers; anthers usually glabrous or shortly hairy at the line of fusion. **Host**: almost exclusively on *Centaurea scabiosa* (Asteraceae), rarely on other *Centaurea* species. **Ecology and distribution**: semi-xerothermic grasslands, grassy slopes, dry meadows, field edges and fallow land; restricted to chalk in the north of its range. Flowering rather late, mid-June–July (rarely August). Western and central Europe, Nordic and Baltic countries; a temperate-European sub-Atlantic floristic element. **Key diagnostics**: stout and pale yellow-brown spikes with many, dense flowers; scales usually long and numerous; broadly tubular and regularly curved corolla with divergent elliptic lower lip lobes. **Similar species**: *Orobanche elatior*, *O. centaurina*, *O. leptantha* and *O. loscosii* are all considered by some authors to be conspecific, or distinct entities based on various morphological and ecological characters. *Orobanche centaurina* differs in its carrot-red to whitish-rosaceous colouration, shorter scales, dorsally straight flowers showing a flattened adaxial surface and big, divergent, spathulate, concave lower-lip lobes and patent, wide upper-lip lobes; *O. leptantha* differs in having rounded, convergent lower lip lobes and longer bracts, often long exceeding the flowers, the calyx free, contiguous or rarely connate, subentire or with unequally bidentate calyx segments; *O. loscosii* differs mainly in its less patent flowers, olive-brown stigma and lower filament insertion. *Orobanche elatior* is also morphologically close to the *O. alsatica* agg., and could be confused with *O. flava*, however the diagnostics stressed above, and host differences, are distinct.

Orobanche elatior
UK (right, page 335)
Poland (page 334)

NAME ***Orobanche flava*** Mart. ex F.W.Schultz
SYNONYMS ***Orobanche tussilaginis*** Mutel; ***Orobanche petasitis*** Fisch.

Habit: stout, often clumped; inflorescence cylindrical, many-flowered, usually ± lax, rarely ± dense. **Height**: (15–)40–70(–80) cm. **Stem**: stout, erect, orange-yellow, yellowish, brownish, purplish, glandular-pubescent; scales triangular, elongated or lanceolate, 10–25 mm, scattered not overlapping. **Calyx**: segments entire or unequally bidentate, elongated-acute, shorter than or ½ as long as the corolla; 7–10 mm. Floral bracts lanceolate, acute, as long as corolla or longer, glandular-pubescent. **Corolla**: erect to erecto-patent; tubular, the dorsal line evenly curved from the base onwards, often bent forward, sometimes slightly straightened near the upper lip; (15–)18–20(–25) mm; yellowish, yellow, reddish-brown, orange-brown. Upper lip bilobate, deeply emarginated, pubescent, lobes are porrect and later deflexed; lower lip 3-lobed, middle lobe is usually larger, crenate and sparsely glandular-pubescent. **Carpel**: stigma lobes spherical, orange or yellow; style sparsely glandular-pubescent or glabrous. **Stamens**: filaments inserted 3–6 mm above corolla tube, densely pubescent from the base to the middle and more sparsely pubescent to the anthers; anthers pubescent at the line of fusion. **Host**: parasitic only on *Petasites*, for example *P. albus*, *P. kablikianus*, *P. hybridus* and *P. paradoxus* (Asteraceae). **Ecology and distribution**: tall, humid, herbaceous vegetation, forest and stream edges, river valleys, alpine foothills and subalpine mountains; rarely naturalised in botanic gardens. Flowering (June) July–August, fruiting (July) August. An Alpine–Carpathian species that extends from western Europe to southern, central and eastern Europe to the Caucasus. **Key diagnostics**: clumped, yellowish spikes; filaments inserted 3–6 mm above corolla tube, densely pubescent from the base; normally easily identified by its distinct host and ecology. **Similar species**: *Orobanche flava* is distinctive and difficult to confuse; superficially similar to *O. cicerbitae*, *O. mayeri* and *O. lucorum*, but has different host and ecology; in particular, *O. lucorum* has a porrect upper lip to the corolla and filaments inserted near the corolla base; see also features emphasised above. *Orobanche flava* var. *doriae* Emb. & Maire, which was described from the Grand Atlas mountain range of Morocco, a parasite of the *Senecio doria*, should be considered at the species rank, *O. doriae* (Emb. & Maire) Ó. Sánchez & Piwow. This species is distinct from typical European *O. flava* because it is slenderer, with less dense inflorescences, except at the apex, with erecto-patent flowers with proportionally, narrower tubes, and less conspicuous indumentum in all parts.

Orobanche flava
UK (left, right)
Poland (page 338, 339)

NAME ***Orobanche foetida*** **Poir.**
SYNONYM ***Orobanche fusca*** **Lam.**

Habit: typically very robust, many-flowered, congested spikes. **Stem**: 20–70(–100) cm tall and 0.4–1.0 cm wide, simple, swollen at the base and with abundant yellow glandular hairs; scales 1.5–4.0 × 0.3–0.5 cm, red-brown with glandular hairs. **Calyx**: ca. 1.5 cm, with acuminate segments of similar length and glandular hairs. Floral bracts 1.5–3.0 cm, lanceolate and red-brown with glandular hairs, often much-exceeding the flowers, especially in bud. **Corolla**: 2.0–2.5 cm, erecto-patent, sub-campanulate, sometimes abaxially curved; exterior red-brown to blackish, with dispersed glandular hairs; interior shining dark chestnut-brown; upper lip bilobed; lower lip subequally trilobed with lobes rounded and often finely toothed. **Carpel**: stigma bright yellow (rarely purple), with diverging lobes. **Stamens**: inserted markedly above (3–)3.5(–7) mm above) the base of the corolla, filaments with glandular hairs above and non-glandular hairs below and towards the base; anthers glabrous. **Host**: a variety of herbaceous and sub-shrubby Fabaceae, especially *Ononis* and *Scorpiurus*, as well as *Lotus*, *Lathyrus*, *Medicago*, *Vicia* and *Trifolium*. **Ecology and distribution**: Iberia, Balearic Islands and North Africa, extending east to Libya. Often coastal – on fixed dunes and similar habitats. Flowering (April) May–June. **Key diagnostics**: typically very robust, congested spikes of flowers with dark chestnut-brown interior with contrasting yellow stigma, foetid scent (not always discernible) and a lower lip divided into subequal lobes with finely toothed margins; filaments inserted 3–7 mm above the corolla base. **Similar species**: *Orobanche gracilis* has a sparser inflorescence, fragrant flowers with different shape, fused stigma lobes, and ciliate lower lip margins. *Orobanche variegata* also has a foetid scent but a diagnostic large deflexed middle lobe on the lower corolla lip; both of these species also possess a shiny, dark reddish inner corolla surface. All-yellow forms of *O. foetida* have been confused with *O. densiflora*, which is similar but typically whitish-cream, possess a different (not foetid) fragrance, have hairy filaments with a lower point of insertion and with broadened, deltate bases, and are parasitic mainly on *Lotus creticus*. Molecular work by the authors (unpublished) revealed that *O. foetida* in Portugal is genetically distinct from populations of *O. densiflora* in Cadiz in Spain, despite morphological and ecological similarities.

Orobanche foetida
Portugal (left, right, page 342, 343)

NAME ***Orobanche gamosepala*** Reut.
SYNONYMS ***Ceratocalyx reuteri*** Lange; ***Necranthus orobanchoides*** Gilli

Habit: slender to slightly stout. **Height**: 5–45 cm. **Stem**: slender, dark reddish-brown or violet-brown, thickened at base, with few oblong-ovate, up to 20 mm long scales, glandular-pubescent, with sometimes longer white hairs. Inflorescence cylindrical or ovoid, usually few-flowered and shorter than remaining stem. **Calyx**: gamosepalous, segments connate (cup-shaped), broad, bidentate, not subulate at tip, rarely not in adaxial side, with 2–4 lanceolate teeth, glandular and ciliate; ⅓ or rarely ½ of the corolla. Floral bracts ovate, acuminate, usually nearly equalling calyx, copiously ciliate-pilose. **Corolla**: upright, tubular-campanulate, the dorsal line erect for a considerable extent; 25–35 mm long; usually dark red, deep violet, black-purplish-violet, or rarely brownish-red (black or nearly so when dry), glandular-puberulent. Upper lip barely sinuate, carinate, hooded, galeate (helmet-shaped), with 2 ± ovoid to truncate, rounded lobes, glandular-pubescent; lower lip 3-lobed, lobes suborbicular or obtuse, short, subequal, with equal margins, deflexed, glandular-pubescent. **Carpel**: stigma lobes yellow to yellow-orange, touching; style sub-glabrous. **Stamens**: filaments inserted 3–7 mm above base of corolla tube; sub-glabrous or shortly glandular pilose below stigma; anthers ± hairy. **Host**: exclusively *Geranium* spp. (Geraniaceae), for example *G. gracile*, *G. gymnocaulon*, *G. ibericum*, *G. psilostemon* and *G. sylvaticum*. **Ecology and distribution**: mainly subalpine and alpine meadows, rocky, grassy slopes, screes, scrub, forests, usually (1,700–)2,000–2,600(–2,900) m. Flowering (end May) June–July (August), fruiting end July–August. Endemic to the Caucasus; mainly in Ciscaucasia (European Russia in Greater Caucasus), less abundant in Transcaucasia (Georgia, Armenia, Azerbaijan and north-east Turkey). **Key diagnostics**: blackish spikes of flowers (black also in herbarium specimens), with contrasting yellow stigmas; inflorescence few-flowered, flowers erect, corolla galeate and its dorsal line straight, gamosepalous; subalpine or alpine habitats, and host. **Similar species**: none. Phylogenetically, this species forms one of the oldest evolutionary clades together with *O. colorata*.

Orobanche gamosepala
Georgia (right, page 346, 347)

NAME ***Orobanche gigantea*** (Beck) Gontsch.

SYNONYMS ***Orobanche kotschyi* var. *gigantea*** Beck; ***Orobanche pamiroalaica*** Gontsch. & Novopokr.

Habit: robust; inflorescence cylindrical, oblong or ovate, usually dense and many-flowered, rarely few-flowered, as long as its stem. **Height**: up to 70 cm. **Stem**: robust, stout, clavate at the base, yellowish, glandular-pubescent with whitish hairs; scales lanceolate, up to 25 mm long. **Calyx**: segments free, in the middle divided into 2 lanceolate-subulate teeth, rarely short bidentate or entire, 12–20 mm. Floral bracts up to 22 mm long, lanceolate, white pubescent or lanate. **Corolla**: usually patent, tubular, strongly geniculate, slightly bent forward to almost straight, without a noticeable constriction, often distinctly patent towards the limb; 25–40 mm; brown-yellow, yellowish, yellowish-violet, yellowish-red, often cream-brown at the base with purple-blue or brown limbs, glandular-hairy outside, veined; upper lip 2-lobed, lobes broad and deflexed; lower lip 3-lobed, lobes broad, ovate, deflexed, middle lobe usually slightly longer. **Carpel**: stigma yellow, conspicuous; style rarely with glandular hairs. **Stamens**: filaments inserted 10–14 mm above corolla base, hairy below, glabrous above, glandular-pubescent below the anthers; anthers hairy at base. **Host**: *Ferula*, possibly other Apiaceae. **Ecology and distribution**: foothills, lower and middle mountain belts, steppe, rocky slopes and semi-deserts. Flowering usually (end April) May–July. Central Asia: Kazakhstan, Kyrgyzstan, Tajikistan and Uzbekistan. **Key diagnostics**: a polymorphic, usually robust plant, up to 70 cm, with dense inflorescence and stem clavate at base, with large, 25–40 mm long, tubular, strongly geniculate and patent flowers, and high filament insertion. **Similar species**: similar to the poorly known and variable *O. kotschyi* and probably also *O. sordida* C.A.Mey., both of which are also reported to grow on Apiaceae. *Orobanche kotschyi* differs in having shorter scales and shorter corolla (18–25 mm), with a distinct constriction, slightly patent in limb, and a lower filament insertion, 7–10 mm. It can be somewhat similar to *O. sulphurea* Gontsch., which is cited to grow on *Crambe* (Brassicaceae), and to *O. solenanthi* Novopokr. & Pissjauk., reported on *Solenanthus* (Boraginaceae); both are described from Tajikistan; each of the species cited above require further investigation, especially from central Asia to Iran. Plants identified as *O. gigantea* f. *talassica* Tzvel. from western Thien Shan in Kazakhstan have ovate, sometimes almost capitate, and few-flowered inflorescences with almost straight flowers; these are somewhat similar to *O. amoena* f. *colossea*, which has a bluish-violet, erecto-patent corolla that is straight or slightly geniculate towards the middle, white stigma, and rather sparse inflorescence; it is generally a smaller plant without a conspicuously clavate stem base, and is generally parasitic on *Artemisia*. *Orobanche rapum-genistae* differs in the markedly distant stigma lobes, hooded upper corolla lip, stamens inserted at the corolla base, ecology and distribution.

Orobanche gigantea
Kazakhstan (right, page 350, 351)
photos: Aleksandr Ebel

NAME ***Orobanche gracilis*** **Sm.**
SYNONYMS ***Orobanche cruenta*** **subsp. *gracilis*** (Sm.) Arcang.; ***Orobanche cruenta*** Bertol.

Habit: slender stems with rather large-flowered, lax inflorescences; sometimes clumped. **Stem**: height 15–60 cm, width 0.2–0.7 cm, simple, slightly swollen at the base, reddish-yellow with abundant glandular hairs; scales 0.7–1.5 cm, ovate-deltate to linear-lanceolate, glandular-pubescent. **Calyx**: segments unequal, free or connate at the base, ending in narrow teeth, yellowish-red with glandular hairs; 0.8–0.15 cm. Floral bracts 1.0–2.0 cm, linear-lanceolate and reddish, with glandular hairs. **Corolla**: tubular-campanulate, erecto-patent, slightly curved abaxially with some constriction at the middle; glandular-pubescent with shining dark red interior and yellow exterior surface with red venation towards the lip; 1.5–2.5 cm. Upper lip somewhat bilobed, emarginate; lower lip subequally trilobed. Lobes somewhat rounded, with ciliate margins; flowers fragrant. **Carpel**: stigma bright yellow, with fused lobes. **Stamens**: filaments inserted 1.5–3.5 mm above base of the corolla, with glandular hairs distally and non-glandular hairs towards the base; anthers glabrous. **Host**: parasitic on various Fabaceae, especially *Genista*, *Cytisus*, *Lotus*, rarely *Anthyllis* and *Dorycnium*. **Ecology and distribution**: widespread and common in various habitats including garrigue, high elevation grassy habitats, xerothermic grasslands, open woods and roadsides. From north-west Africa, across western, southern and central Europe to Ukraine, Turkey and the Caucasus. **Key diagnostics**: shiny dark red corolla interior paired with fragrant scent, fused stigma lobes, ciliate lower lip margins, and stamen insertion point usually <3 mm from corolla base. **Similar species**: *Orobanche foetida* has filaments inserted more than 3 mm above the base of the corolla, and stigma lobes spreading; *O. variegata* has a foetid scent and a large, deflexed middle lobe on the lower lip of the corolla. *Orobanche austrohispanica* is very similar (also considered conspecific) and differentiated by its ± dorsally keeled, distinctively and contrastingly reddish-veined, pale corollas that lack the shiny red interior of *O. gracilis*, and in its almost glabrous filaments and filament bases.

Orobanche gracilis
Montenegro (left)
Portugal (right)
Montenegro (page 354)
Georgia (page 355)

NAME ***Orobanche grenieri*** F.W.Schultz
SYNONYM ***Orobanche lactucae*** Arv.-Touv.

Habit: slender with dense inflorescence shorter than its stem. **Height**: 16–38(–52) cm. **Stem**: usually ± slender, yellowish, yellowish-brown, reddish-brown, glandular-pubescent; scales lanceolate to narrowly lanceolate, lax. **Calyx**: segments 2-toothed or entire, teeth acuminate and ± filiform to apex, 7–11 mm long. Floral bracts ovate-lanceolate to lanceolate, acuminate. **Corolla**: erect to erecto-patent, tubular to tubular-campanulate; dorsal line of the corolla geniculate in the lower part and ± almost straight in the rest; (15–)17–18 mm; bluish-violet, basally yellowish-white, veined, with usually whitish mouth; outside sparsely, short glandular-pubescent, inside almost glabrous. Upper lip with 2 broad lobes, usually glabrous; lower lip with 3 usually glabrous, broad, subequal, obovate lobes; folds white. **Carpel**: stigma yellowish or whitish; style sub-glabrous. **Stamens**: filaments and anthers glabrous. **Host**: Asteraceae, mainly *Lactuca* spp., e.g. *L. viminea*, *L. takhtadzhianii*, *L. quercina* subsp. *wilhemsiana*, *L. serriola* and occasionally *Tragopogon graminifolius*. **Ecology and distribution**: roadsides, rocky, loamy slopes, edges of fields, semi-deserts, usually at (500–)900–1,400(–1,700) m. Flowering (end of May)–June (July), fruiting July–August. South-western Europe (Spain, France and Italy), and recently also reported in eastern Europe (Russia and Crimea), western and central Asia (Georgia, Armenia, Azerbaijan, Iran, Tajikistan, Kyrgyzstan and Turkey). **Key diagnostics**: dorsal line of the bluish-violet corolla geniculate in the lower part and ± almost straight in the rest, broad lobes of the corolla; host-specific. **Similar species**: *Orobanche grenieri* is closely related to *O. cernua* and *O. cumana*: the former differs from these species in its slenderer stem with a short inflorescence, darker flowers, narrow bracts, longer corolla with dorsal line geniculate in the proximal part and ± almost straight in the distal part, and large and broad lobes. These species have different hosts and ecologies.

Orobanche grenieri
Georgia (left, right)

NAME ***Orobanche grossheimii*** Novopokr.

Habit: robust, sometimes clump-forming; inflorescence equal to or longer than the remaining stem, rarely shorter, many-flowered, usually ± dense, and rarely ± lax. **Height**: (35–)40–50(–60) cm. **Stem**: dull pink, whitish, reddish, rosaceous, pale yellow or yellow-brown; stout, densely glandular-pubescent usually on the whole part, with reddish to violet, or rarely pale yellowish glandular hairs; scales (13–)15–25(–30) mm, usually dense, broadly lanceolate-ovate to triangular and lanceolate. Inflorescence (19–)20–25(–27) cm, sub-cylindrical to cylindrical, tapering to the apex. **Calyx**: segments free, asymmetrically ovate and unequally bidentate or rarely entire, with 2 long narrowly lanceolate calyx-teeth, usually longer than ½ the corolla tube, (11–)17–20(–22) mm long. Floral bracts (22–)25–31(–36) mm, equal or longer, rarely shorter than the corolla, broad lanceolate. **Corolla**: ± erecto-patent and large-sized, tubular to narrowly campanulate (considerably widened at the mouth); slightly inflated at the filament insertion point on the lower side; dorsal line curved at base, then almost straight; (16–)24–26(–30) mm long, (7–)8–9(–10) mm in diameter in the central part; pale pinkish-white, yellow, brown, often darker on the upper part, with darker veins; externally glandular-pubescent, with reddish to violet, rarely pale yellow glandular hairs; interior glabrous but with sparse short, reddish, rarely pale yellowish hairs on the lips. Upper lip with 2 broad lobes, emarginate, sparsely glandular-pubescent also in the inner part; lower lip with 3 suborbicular lobes, irregularly dentate on the margins, the middle lobe usually larger. **Carpel**: stigma lobes elongated-spherical to ovoid, yellow or bright yellow; style with short and glandular hairs throughout. **Stamens**: filaments obliquely inserted at 3–4 mm above the corolla base, densely hairy in the lower ½, not glandular, densely glandular-pubescent below the anthers; anthers short-pubescent along sutures of thecae, especially in the ¾ part. **Host**: *Cephalaria* spp. (Caprifoliaceae), especially *C. gigantea*, rarely *C. tchihatchewii*. **Ecology and distribution**: meadows, grassy and stony slopes from middle mountain belts up to alpine belt, pastures, rarely forests, usually (1,800–)2,000–2,700(–3,000) m. Flowering (end of May) June–July (August), fruiting July–August. Endemic to the Caucasus, in Russia, Georgia, Armenia and Azerbaijan. **Key diagnostics**: robust, with numerous, almost overlapping scales; inflorescence usually ± dense, corolla yellowish, yellowish-brown, whitish ± tinged with purple, paler inside (light yellow or whitish); parasitic only on Caprifoliaceae (*Cephalaria*). **Similar species**: *Orobanche kurdica* is similar but with an inflorescence usually ± lax, longer corolla, bracts equal or shorter, stamens with a higher insertion point; *O. schelkovnikovii* is similar but has a stem with few scales, calyx segments conspicuously connate on abaxial side, and a different host range.

Orobanche grossheimii
Armenia (right)

NAME ***Orobanche haenseleri*** Reut.

Habit: slightly robust; often clumped; inflorescence 5–25 cm, relatively dense, many-flowered. **Height**: 20–40(–70) cm. **Stem**: robust, yellowish-pink, orange-brown, with abundant glandular hairs; scales 12–20 mm, lanceolate, similar in colour to the stem but darkening when mature, with glandular hairs. **Calyx**: 13–20 mm, often equalling the corolla tube with segments unequally divided into 2 broadly lanceolate teeth, pink, sometimes yellowish, with glandular hairs. Floral bracts 16–24 mm, linear-lanceolate, ranging in colour from pink to brown, chestnut, and darkening with age, with glandular hairs. **Corolla**: 20–25(–28) mm, from sub-erect to ± patent, tubular-campanulate, slightly infundibuliform, straight in the basal ⅔, curved towards the apex, dark pink with somewhat darker veins, dull orange-red, reddish orange, pale salmon to wine-coloured, both externally and internally, with abundant glandular hairs on the outer surface. Upper lip somewhat bilobed; lower lip lobes clearly divergent, the central lobe larger than the lateral ones, deflexed, lobes oblong or obovate, not ciliated or serrated at the margin, rarely irregularly toothed. **Carpel**: stigma with markedly divergent lobes, bright yellow; style with scattered short glandular hairs, somewhat denser in the upper ½, pale yellow, often tinged pale pink, sometimes purple. **Stamens**: filaments inserted 3–4 mm above the base of the corolla, somewhat widened at their base, with glandular and non-glandular hairs along their entire length; anthers glabrous. **Host**: Ranunculaceae, especially *Helleborus foetidus*, rarely *H. viridis*; occasionally reported on Lamiaceae (*Sideritis* and *Phlomis*). **Ecology and distribution**: usually montane habitats, from the French Pyrenees to southern Spain in coniferous and mixed forests, forest clearings, mountain slopes and scree, river and stream gorges, at (700–)1,000–2,450 m. Flowering May–July (August). **Key diagnostics**: corolla with an open throat, reddish-orange, pale salmon to wine-coloured, both externally and internally with a contrasting yellow stigma; filaments hairy along their entire length; host-specific. **Similar species**: somewhat similar to *O. rapum-genistae*, mainly owing to its robust, clumped habit, but the flower morphology and host are distinct; phylogenetically closely related to *O. ebuli*, *O. lucorum*, *O. salviae* and *O. mayeri*, differing in the characters stressed above.

Orobanche haenseleri
Spain (left, right)
photos: Gonzalo Moreno Moral
Spain (page 362, 363)
photos: Mario Mairal Pisa

NAME ***Orobanche hederae*** Duby
SYNONYMS ***Orobanche laurina*** Bonap. ex Bertol.; ***Orobanche vaucheri*** Noulet

Habit: slender to robust stems with flowers lax, from about ground-level, becoming denser above. **Height**: 20–35 cm. **Stem**: 0.5–0.8 cm across, strongly swollen at the base, and with glandular hairs, reddish to mauve-purple; scales 1.8–2.5 cm, widely lanceolate and glandular-pubescent. **Calyx**: 10–15 mm, bifid or entire. Floral bracts 12–22 mm, equalling or exceeding the flowers, ovate-lanceolate to lanceolate, acuminate and minutely toothed near the tip. **Corolla**: dull cream, tinged reddish-purple; sparsely glandular or almost hairless, erecto-patent to patent tube slightly inflated below, usually constricted distally; upper lip entire to emarginate, the lobes forwardly directed or spreading; (10–)12–20(–22) mm; lower lip with middle lobe just the largest, all subacute, crisped and toothed; corolla rather straight along the back, except at the base. **Carpel**: stigma lobes partially united, yellow (sometimes tinted pink or reddish-orange). **Stamens**: inserted 0.3–0.4 cm above the base of the corolla, filaments glabrous distally and slightly hairy near the base; anthers glabrous. **Host**: almost exclusively parasitic on *Hedera* spp. (Araliaceae); rarely on *Fatsia*, *Tetrapanax* or *Eryngium* (Apiaceae) in gardens. **Ecology and distribution**: open woods, cliffs, crags and gorges, often coastal; increasingly frequent in towns and gardens in the north of its range. Widespread and common across Europe, north to the UK, Germany, Ukraine, and south-east to Turkey, the Caucasus and Iran; also in Mediterranean Africa and Macaronesia. **Key diagnostics**: distinguished easily by its habitat (typically rather shady areas with abundant ivy), its long purple spikes with flowers almost throughout, i.e. from near ground-level, and (usually) yellow stigma. Corolla narrow and pinched towards the tip, the lobes rather acute and with somewhat lacerated edges. The species has the longest flowering period, sometimes even flowering at new year. **Similar species**: mistaken frequently for *O. minor* (which can also grow on *Hedera*, but rarely), from which it can be distinguished most readily by its yellow stigma lobes (in *O. minor* usually pinkish, and yellow only in all-yellow forms). It also has a different corolla shape (somewhat dorso-ventrally constricted behind the mouth), flowers extending over most of the stem, and a broad, pointed top to the spike.

Orobanche hederae
UK (left, right, page 366, 367)

NAME ***Orobanche iammonensis*** A.Pujadas & P.Fraga

Habit: short, stout. **Height**: (6–)10–15(–17) cm high. **Stem**: simple, 0.3–0.6 cm diameter in middle part, becoming gradually and uniformly slightly swollen at the base; bulbous at the base, 1.3–1.7 cm in diameter; slightly striated, densely glandular-pubescent; pale yellow. **Calyx**: shorter than or equalling the corolla tube, segments simple, entire or subentire, narrowing to the apex and long acuminate, or unequally bifid with a wide ovate base and long acuminate subulate teeth; segments free, clearly separate; 0.7–1.0 cm. Floral bracts 1.0–1.3 cm, equalling the corolla, narrowly lanceolate, deflexed, with yellow glandular hairs. **Corolla**: narrowly tubular, only slightly inflated at base, uniformly curved, sometimes geniculate, erecto-patent to sub-patent, externally glandular-pubescent with yellow glandular hairs; 1.0–1.4 cm long, 0.3–0.4 cm in diameter in centre; pale yellow. Upper lip emarginate, erect or patent; lower lip with central lobe slightly larger than the laterals, irregularly dentate on margins. **Carpel**: stigma lobes pale yellow; style with sparse glandular hairs, denser in the lower part and beneath the stigma. **Stamens**: filaments obliquely inserted, geniculate, white, nectaries at filament base yellow; anthers 1.5 mm, ochre, paler when dry. **Host**: *Anthemis maritima* (Asteraceae). **Ecology and distribution**: known only from the type locality in north-western Menorca, Balearic Islands, on coastal rocky slopes. **Key diagnostics**: small yellow plants; geniculate filaments; host-specific. **Similar species**: belonging to the complex Subsection *Minores*: similar to yellow forms of *O. minor*, but smaller with subulate calyx lobes; corolla smaller and more strongly curved to geniculate, the base densely hairy and with non-glandular hairs (*O. minor* has only glandular hairs); stamen filaments geniculate (straight or slightly curved in *O. minor*). Also similar to *O. litorea*, but *O. iammonensis* is a smaller plant; the lower part of the stem is bulbous (slender and elongated in *O. litorea*); the corolla of 1.0–1.4 cm is also smaller (1.3–1.7 cm in *O. litorea*).

Orobanche iammonensis Menorca, Spain (left, right) photos: Pere Fraga

NAME ***Orobanche ingens*** (Beck) Tzvelev
SYNONYM ***Orobanche alba* f. *ingens*** Beck

Habit: robust, rarely slender. **Height**: (35–)60–100 cm. **Stem**: usually stout, purple-brownish, yellowish-brown, reddish-brown, glandular-pubescent with long stiff hairs; scales lax, lanceolate. Inflorescence usually long, rarely short, cylindrical, somewhat lax, usually many-flowered. **Calyx**: segments deeply bidentate, teeth frequently filiform, 10–16 mm. Floral bracts as long as the corolla tube, slightly shorter than the corolla. **Corolla**: campanulate, dorsal line almost straight, except evenly curved at the proximal part; (20–)23–24(–30) mm; yellow, rarely violet, dark pink, veined. Upper lip with 2 shallow, short, broad lobes, emarginated, erect or rarely patent; lower lip 3-lobed, lobes ovate, irregularly dentate on the margins. **Carpel**: stigma yellow or orange, rarely reddish, violet, dark pink; style with glandular hairs, denser below stigma. **Stamens**: filaments inserted 1(–2) mm above the corolla base; with long hairs on the basal ½ and sparsely glandular-pubescent (almost glabrous) under the anthers. **Host**: *Heracleum* spp., including *H. leskovii*, *H. sosnowskyi* and *H. mantegazzianum*; possibly on other large Apiaceae, such as *Ligusticum alatum*. **Ecology and distribution**: in tall herbaceous vegetation, mainly in subalpine meadows, pastures, forest edges and glades, near river valleys and fallow land, usually at (1,000–)2,000–2,300(–2,500) m. Flowering (June) July–August, fruiting (July) August (September). Endemic to the Caucasus, mostly in the Greater Caucasus range, in Russia and Georgia. Possibly also in Azerbaijan. **Key diagnostics**: usually tall with a long inflorescence, campanulate corolla with a mixture of longer and stiff whitish hairs, deeply bidentate calyx, and very low filament insertion point; host-specific. **Similar species**: a distinctive species, somewhat similar to *O. laxissima*, and to lesser extent, to *O. bartlingii* or *O. inulae*, which differ in corolla size and shape (tubular to tubular-infundibuliform), calyx characteristics and host range.

Orobanche ingens
Georgia (right, page 372, 373)

NAME ***Orobanche inulae*** Novopokr. & Abramov

Habit: somewhat lightly robust, rarely slender; inflorescence short cylindrical or oblong, ± dense above, sometimes lax below, usually shorter than its stem. **Height**: (5–)15–30(–34) cm. **Stem**: slender to stout, yellow or yellowish-brown with few lanceolate scales, up to 15 mm long. **Calyx**: segments free, entire (rarely 2-toothed), lanceolate to narrowly lanceolate and almost filiform at the apex; (8–)10–15 × (1.9–)3–4 mm. Floral bracts 13–16(–18) mm, ± shorter than the corolla, ovate-lanceolate. **Corolla**: broadly tubular, tubular-campanulate; dorsal line regularly arched-curved; (15–)17–20 × 4.5–6.0 mm; pale, brownish-yellow, light yellow. Upper lip evenly curved, emarginate or a little bilobed, porrect at first, later slightly patent; lower lip ± equal 3-lobed, lobes large, ovate-rounded or angular, with crenate to irregularly dentate, glandular margins. **Carpel**: stigma with broadly connected subspherical yellow lobes; style glandular-hairy or sub-glabrous throughout. **Stamens**: filaments inserted (1.5–)2–3 mm; in the lower ½ to ⅔ of their length long-hairy, almost glabrous above or glabrous throughout, sometimes with scattered, short glandular hairs below the anthers; anthers glabrous or hairy at suture. **Host**: *Inula* spp., usually *I. orientalis* (Asteraceae). **Ecology and distribution**: subalpine meadows, tall herbaceous vegetation, rocky slopes, forests and scrub, and forest glades, usually 2,000–2,500 m. Flowering July (August), fruiting end July–August. Caucasian endemic. Greater Caucasus in Russia, Georgia and Azerbaijan, and also reported from Lesser Caucasus in Armenia. **Key diagnostics**: yellow or brownish-yellow corolla with dorsal line regularly arched-curved; stamens inserted (1.5–)2–3 mm above the corolla base; host-specific. **Similar species**: differs from *O. arpica* or *O. cicerbitae* in inflorescence stature, bracts and corolla (corolla with dorsal line regularly arched-curved), lobe shape, filament insertion and host species.

Orobanche inulae
Georgia (left, right)

NAME ***Orobanche javakhetica*** Piwow., Ó.Sánchez & Moreno Mor.

Habit: slender to slightly robust; inflorescence cylindrical to slightly ovoid, shorter than remaining stem, ± dense. **Height**: (20–)25–30(–35) cm. **Stem**: slender, pale white-yellow or rarely pale yellow-brown; scales lanceolate to narrowly lanceolate. **Calyx**: segments, free, entire, narrow, with teeth long, acuminate almost filiform at apex and with bases ovate, pale pink, pale yellow-brown, dense glandular-hairy; (7–)8–9(–11) mm long, shorter than ½ the width of the corolla tube. Floral bracts (11–)13–14(–16) mm, shorter or rarely equalling the corolla, broad ovate-lanceolate, with dense white or pale yellow glandular hairs, darker than the stem scales, pinkish-brown, violet-brown. **Corolla**: erecto-patent, tubular to narrowly campanulate (slightly widened at the mouth), usually strongly inflated above the filament insertion, and strongly constricted below; (15–)18–19(–21) mm long, (5–)6–7(–8) mm in diameter in the central part; pale yellow-pink, white-yellow, yellowish-white, frequently pinkish at the apex (lobes), with dirty white, pale pink or pale yellow veins. Upper lip slightly emarginate, with 2 short broad lobes, usually facing downward, sparsely glandular-pubescent also in the inner part; lower lip with 3 ovate, or ± rectangular lobes, central lobe slightly larger than laterals; lobes irregularly dentate on margins, sparsely glandular-pubescent also in the inner part. **Carpel**: stigma lobes elongated-spherical to ovate, yellow or bright yellow; style glabrous or with sparse short glandular hairs below stigma. **Stamens**: filaments obliquely inserted at 3.5–4.5 mm above the corolla base, densely villous (not glandular) from base to ± middle part; upper part densely glandular-pubescent; anthers basally and along ¾ of suture distinctly hairy. **Host**: *Stachys macrantha* (Lamiaceae), rarely *Lomelosia caucasica* (Caprifoliaceae). **Ecology and distribution**: subalpine meadows, rocky slopes, 2,200–2,600 m. Flowering in second half of July, fruiting in August. Probably endemic to the Caucasus, known from Armenia and Georgia. **Key diagnostics**: pale flowers with pinkish lobes; corolla tubular to narrowly campanulate; dorsal line ± flexed forwards or straight at the apex; floral bracts clearly darker than the stem scales and flowers, pinkish-brown, violet-brown; host-specific. **Similar species**: most similar to *O. inulae* and *O. arpica*; distinguished from both by the characters emphasised above.

Orobanche javakhetica
Armenia (left, right)

NAME ***Orobanche kotschyi*** Reut.

SYNONYMS ***Orobanche kotschyi* var. *typica*** Beck; ***Orobanche spectabilis*** Reut.; ***Orobanche xanthostachya*** Novopokr.; ***Orobanche bucharica*** Novopokr.

Habit: robust; inflorescence cylindrical or oblong, dense and many-flowered, ± equalling its stem; glandular-pubescent, usually with a mixture of longer whitish hairs or wholly white arachnoid-lanate around the inflorescence. **Height**: up to 70 cm. **Stem**: robust, yellowish, yellowish-brown, 5–15 mm thick, clavate at the base; scales lanceolate or ovate-lanceolate, up to 10 mm long. **Calyx**: segments free, 12–15 mm long, light brown with darker veins, usually dissected beyond the middle into 2 lanceolate-subulate teeth, rarely short-bidentate. Floral bracts ovate-lanceolate or broadly lanceolate, longer than or equalling the calyx, usually whitish-lanate. **Corolla**: tubular, 18–25(–30) mm long, slightly forward-bent, with a distinct constriction above the stamen insertion point, slightly patent in the limb; brownish-yellow, whitish at the base, rarely with a pale blue or purple tinge in the limb, glandular-hairy externally. **Carpel**: stigma 2-lobed, yellow; style sub-glabrous. **Stamens**: filaments inserted 7–10 mm above base of the corolla tube, sparsely hairy at the base, glabrous above; anthers pubescent at the base. **Host**: *Ferula*, *Prangos* and *Zosima* (Apiaceae); full host range needs verification. **Ecology and distribution**: a rare plant of rocky slopes, the range of which is poorly known. Iran (type locality) through central Asia: Kazakhstan, Turkmenistan, Afghanistan, Tajikistan, Pakistan, Uzbekistan, Kyrgyzstan and China (Xinjang). Flowering April–June (July). **Key diagnostics**: a polymorphic, usually yellowish-brownish, robust plant with congested, many-flowered inflorescences; corolla large, with a distinct constriction and slightly patent in limb; yellow stigmas rather large and conspicuous. **Similar species**: a variable plant for which relationships with other species remain unclear: *O. gigantea* (also reported on Apiaceae) is sometimes included under *O. kotschyi* (*O. kotschyi* var. *gigantea*); both taxa require investigation. *Orobanche gigantea* differs in having longer scales (up to 25 mm), longer corolla (25–40 mm) with a less noticeable constriction and often distinctly patent towards the limb and higher filament insertion (10–14 mm). Possibly also somewhat similar to the poorly known *O. sordida*.

Orobanche kotschyi Kazakhstan (above, right, page 380, 381) photos: Vladimir Kolbintsev

NAME ***Orobanche krylowii*** Beck
SYNONYMS ***Orobanche major* f. *krylowii*** (Beck) Beck; ***Orobanche major* subsp. *krylowii*** (Beck) Nyman

Habit: slender or somewhat robust; inflorescence (4.0–)5–8(–12) mm long, usually with fewer than 20 flowers, cylindrical or oblong, rather lax and with flowers limited to the upper ⅓ of the stem. **Height**: (20–)23–30(–45) cm. **Stem**: brownish-yellow, glandular-pubescent, somewhat thickened at the base; scales elongated, triangular-lanceolate, (10–)16–18.5(–20) cm long, sub-glabrous or glandular-hairy. **Calyx**: 6–12 × 2.4–4.8 mm, segments free, usually entire (with 1 pair of shortly subulate teeth strongly reduced or absent), ovate-lanceolate to subulate. Floral bracts (12.7–)14–17(–20) mm long, shorter than or equalling the corolla. **Corolla**: 17–21 × 5–6 mm, tubular, strongly constricted below the insertion of the stamens, light yellow or whitish, dorsal line gently and regularly curved (but straight near the apex). Upper lip slightly emarginate, porrect; lower lip 3-lobed, with lobes divergent, spathulate, concave and with finely irregularly denticulate margins. **Carpel**: stigma bilobed, with a flattened distal surface, widely bilobed to disciform, bright yellow; style glabrous or nearly on the lower ½ and with inconspicuous, very short glandular hairs on the upper ½. **Stamens**: filaments inserted 2.2–3.5 mm, with abundant, long, glandular hairs and short glandular hairs below the anthers; anthers with short hairs along the sutures. **Host**: Ranunculaceae: *Thalictrum minus*. **Ecology and distribution**: usually in forests, forest clearings, sparse mixed and coniferous woods, and scrub. Flowering June–August. Distribution: south-east Europe (Albania) and eastern Europe (Russia) to Asia: Russia, Kazakhstan, Kyrgyzstan and Mongolia. **Key diagnostics**: usually a bright, whitish to yellowish forest-dwelling plant; corolla tubular with gently and regularly curved dorsal line (straight near the apex); lobes with finely denticulate margins; host-specific. **Similar species**: possibly confused with other Ranunculaceae parasites, such as *O. lycoctoni* and *O. mlokosiewiczii*. Its sister species *O. lycoctoni* occurs from the Cantabrian Mountains in northern Spain to the Eastern Alps, and parasitises *Aconitum lycoctonum*; it is more robust in all parts, has stouter stems, denser inflorescences, and patent and widely tubular flowers with a more marked corolla dorsal geniculation. The Caucasian species *O. mlokosiewiczii* parasitises *A. cymbulatum*, and differs in having larger flowers with campanulate corolla, different lobe morphology and a higher filament insertion. *Orobanche cicerbitae* is another Caucasian species, which differs in its strongly curved corolla, distinct lobe morphology and specificity for Asteraceae.

Orobanche krylowii
Krasnoyarskii Krai, Russia (right)
photo: Petr Kosachev

NAME ***Orobanche kurdica*** Boiss. & Hausskn.
SYNONYM ***Orobanche rosea*** Tzvelev

Habit: robust, with inflorescence (12–)20–28(–40) cm, sub-cylindrical to cylindrical, tapering to the apex, usually equal to or longer than its stem, many-flowered, usually ± lax, and rarely ± dense. **Height**: (25–)45–50(–60) cm. **Stem**: dull pink, reddish, rosaceous, pale yellow or yellow-brown; stout, rarely slender, glandular-pubescent throughout. Scales 20–25(–30) mm long, broadly lanceolate-ovate to triangular and lanceolate. **Calyx**: segments free or rarely connate, asymmetrically ovate and unequally bidentate, with 2 long narrowly lanceolate teeth, longer than ½ the width of the corolla tube; (14–)19–20(–22) mm long. Floral bracts (18–)25–26(–30) mm, equal or shorter, rarely longer than the corolla, broad lanceolate. **Corolla**: large, ± erecto-patent; tubular to narrowly campanulate (considerably widened at the mouth); slightly inflated at the filament insertion point below; dorsal line curved at base, then almost straight; (22–)28–30(–32) mm long, (7–)8–9 mm in diameter in the central part; pale pinkish or yellowish with dull pale pink veins; externally glandular-pubescent; interior glabrous or sparsely hairy on the upper lip. Upper lip sinuate, emarginate, porrect, with 2 broad lobes, these bent backward laterally; lower lip with 3 suborbicular lobes with irregularly dentate margins, the middle lobe usually larger. **Carpel**: stigma lobes elongated-spherical to ovate, yellow or bright yellow; style glandular-pubescent throughout. **Stamens**: filaments obliquely inserted at 4–5(–6) mm above the corolla base; sparsely hairy in the lower ½, not glandular; upper part below the anthers sparsely glandular-pubescent; anthers short-pubescent along sutures of thecae, especially in the ¾ part, rarely glabrous. **Host**: Apiaceae: *Prangos* spp., usually *P. ferulacea*, or less often *P. pabularia* or *Bilacunaria microcarpos*. **Ecology and distribution**: rocky slopes, subalpine meadows, rarely forests, from lower to high mountain belts, (1,200–)1,500–2,700(–3,200) m, in Armenia, Azerbaijan, Georgia, Russia (Dagestan), Iran, Iraq, Turkey and Turkmenistan. **Key diagnostics**: robust, usually rosaceous-yellowish, many-flowered inflorescences with large flowers; corolla curved at base with mouth wide open and deflexed large lobes with dentate margins, calyx segments free. Specific to Apiaceae. **Similar species**: most similar to *O. kochii*, which occurs mainly in central Europe to Crimea. *Orobanche kurdica* is a much larger plant, with longer bracts and inflorescence, calyx segments and corolla, and differs in host and ecology. In the Caucasus, multiple species occur in similar habitats and can be difficult to identify from herbarium specimens: *O. grossheimii* differs in having copious, almost overlapping stem bracts, shorter corolla curved at the base with reddish to violet, rarely pale glands, and free calyx segments and is parasitic only on Caprifoliaceae; *O. schelkovnikovii* has a shorter corolla, slightly curved at the base, lower filament insertion, and calyx segments broadly connate in front, and is parasitic only on Asteraceae.

Orobanche kurdica
Armenia (right, page 386, 387)

NAME ***Orobanche laserpitii-sileris*** Reut. ex Jord.
SYNONYM ***Orobanche alsatica* var. *laserpitii-sileris*** (Reut. ex Jord.) Beck

Habit: usually stout; inflorescence cylindrical or elongated, multi-flowered, dense, or laxer in the lower part, shorter than its stem. **Height**: 30–70(–80) cm. **Stem**: stout, erect, yellowish-brown, brownish, reddish, yellowish, glandular-pubescent; scales dense, triangular to lanceolate. **Calyx**: segments bidentate, rarely entire, oval, free or fused, shorter than ½ the corolla. Floral bracts are lanceolate, acute, as long as or slightly longer than corolla. **Corolla**: 20–30 mm long, widely tubular, densely glandular-pubescent, yellowish-brown, reddish-brown, yellowish at the base, with violet veins, more brightly coloured within, sometimes entire; yellow; corolla dorsal line evenly curved from the base. Upper lip deeply bilobate, lobes are porrect and erect later; lower lip 3-lobed, central lobe larger, all rounded, crenate at the margin. **Carpel**: stigma 2-lobed, lobes spherical, yellow or orange; style glandular-pubescent, especially below stigma. **Stamens**: filaments inserted 5–7 mm above the corolla tube, densely pubescent to the middle, sparsely below the anthers; anthers sub-glabrous. **Host**: exclusively on *Siler montanum* (Apiaceae). **Ecology and distribution**: a central and southern European mountain plant of grassland, screes and stony slopes, at higher elevation. Alps to north-west Balkan Peninsula: Austria, Bosnia and Herzegovina, Bulgaria, Croatia, France, Italy, Montenegro, Serbia, Slovenia, Spain and Switzerland. **Key diagnostics**: a stout plant with dense and multi-flowered inflorescence; flowers dark outside and brightly coloured within; central lobe of lower lip often longer than the laterals; mountain ecology and host. **Similar species**: other species that grow on Apiaceae include *O. alsatica*, *O. bartlingii*, *O. mayeri* and *O. montserratii*; *O. laserpitii-sileris* (typically) has darker and more contrasting pigmentation, a more congested inflorescence with larger flowers, a higher filament insertion, and different corolla and lobe morphology.

Orobanche laserpitii-sileris Croatia (above, right) photos: Marko Doboš

NAME **Orobanche latisquama** (F.W.Schultz) Batt.
SYNONYMS **Boulardia latisquama** F.W.Schultz; **Ceratocalyx macrolepis** Coss.; **Orobanche macrolepis** Pomel

Habit: robust, usually clumped perennial with several stems from the base and long, rather lax, ± glabrous, oblong-ovoid spikes. **Height**: 20–40 cm. **Stem**: glabrous, 4–9 mm in diameter, slightly swollen at the base, pale yellow, pink or deep reddish-purple; scales conspicuous, 2–3 cm, dense (to overlapping) and ovate-oblong, obtuse towards the base and acute towards the inflorescence. **Calyx**: 11–16 mm, entire or bifid, connate in proximal ¼. Floral bracts, exceeding the corolla, very broadly ovate below, abruptly acuminate and blackish apically. **Corolla**: glabrous with subsessile glands, pale yellowish to rose-pink, variably tinged with purple or purple veining towards the apex; 22–35 mm; tube constricted around the middle, inflated below and more campanulate above. Upper lip subentire; lower lip equally trilobed with divergent lobes, margins not ciliate. **Carpel**: stigma pale yellow to whitish, contiguous. **Stamens**: filaments hairy, inserted 7–10(–15) mm above base of corolla; anthers villous. **Hosts**: usually on *Salvia rosmarinus* (Lamiaceae). **Ecology and distribution**: maquis, garrigue and cliffs, especially coastal, across eastern and south-eastern Spain, Portugal, Balearic Islands, Morocco and Algeria. Flowering (March) April–June. **Key diagnostics**: largely glabrous, clumped pinkish (rarely all pale yellow) spikes with numerous conspicuous scales below; filaments inserted over 7 mm above the corolla base. Lower lip without ciliate margins, surface of plant largely glabrous. **Similar species**: a handful of species have similar stature: *O. clausonis* and *O. rapum-genistae* have glabrous anthers and lower filament insertion point in corolla; the former is pale (*O. latisquama* is rarely pale) and the latter has yellow, distant stigma lobes; both differ markedly in their hosts and ecology. Molecular studies suggest these species are divergent (*O. latisquama* interestingly forms a distinct clade that is close to the West Asian *O. gamosepala* and *O. colorata* clade). It is also classified in its own genus under the name *Boulardia latisquama.*

Orobanche latisquama
Spain (left, right)

NAME ***Orobanche laxissima*** Uhlich & Rätzel
SYNONYMS ***Orobanche transcaucasica*** Tzvelev; ***Orobanche benkertii*** Rätzel & Uhlich

Habit: slender, rarely slightly robust, extensively clump-forming, often with numerous spikes; inflorescence shorter than its stem, usually many-flowered, sparse below. **Height**: (10–)25–40(–100) cm. **Stem**: slender to stout, pinkish or brownish, densely glandular-pubescent in the upper part; scales 15–18(–24) mm, ovate-lanceolate to lanceolate. **Calyx**: segments simple or bifid, sometimes with 4 teeth, underdeveloped; segments entire, narrowing to the apex, teeth long and very narrowly acuminate, with a somewhat ovate base; segments free, clearly separate; (6–)13–15(–19) mm long, shorter than the corolla tube. Floral bracts (14–)17(–22) mm, shorter than or equalling the corolla, narrowly lanceolate. **Corolla**: tubular-infundibuliform; the dorsal line slightly curved, sometimes almost horizontal, with umbonate upper lip, only slightly inflated at base; (16–)22–24(–31) mm long, 3–4 mm in diameter in the central part; purple, pink, dull yellow or light brown; externally glandular-pubescent, interior glabrous or scarcely hairy. Upper lip with 2 short, broad lobes, emarginated, erect or rarely patent; lower lip with 3 ovate lobes, the central slightly larger, all with irregularly dentate margins. **Carpel**: stigma lobes rather distant, dark pink, violet to purple, or rarely orange, pale to dark yellow; style with glandular hairs, denser from the stigma side. **Stamens**: filaments inserted at (1.5–)2–3 mm above the corolla base; densely villous at the base or lower ⅓, with non-glandular and glandular hairs, the upper part with short, glandular hairs; anthers glabrous with sparse short glandular hairs located at the line of fusion. **Host**: parasitic on various trees and shrubs including *Fraxinus*, *Ligustrum* (Oleaceae), *Fagus* (Fagaceae), *Carpinus* (Betulaceae), *Acer* (Sapindaceae), *Corylus* (Corylaceae), and occasionally *Punica granatum* (Lythraceae) and *Rhus coriaria* (Anacardiaceae). **Ecology and distribution**: moist, shady or semi-shaded habitats in deciduous forests, parks and gardens, in the submontane and montane zone, usually from 30 m near the Black Sea to (200–)600–1,800(–2,700) m. Flowering (May) June–July, fruiting July–August. Caucasus: Armenia, Azerbaijan, Georgia, Russia and north-east Turkey; needs confirmation in Crimea. **Key diagnostics**: highly polymorphic with variable colouration; inflorescence long, cylindrical, lax, many-flowered; corolla tubular-infundibuliform; often occurs in extensive clumps, in forests. **Similar species**: potentially confused with *O. owerinii*, which has a short-cylindrical to ovoid, ± lax, usually few-flowered inflorescence and calyx segments entire, corolla campanulate-infundibuliform, and higher filament insertion, and occurs in subalpine and alpine habitats such as meadows where it is parasitic on herbs such as *Trifolium*. *Orobanche laxissima* can co-occur with *O. hederae*, which has a smaller, narrowly tubular, corolla that is distinctly constricted below the throat, and grows on *Hedera*. *Orobanche ingens* has a longer inflorescence, campanulate corolla, deeply bidentate calyx, and very low filament insertion point, and grows on *Heracleum*.

Orobanche laxissima
Georgia (right, page 394, 395)

NAME ***Orobanche leptantha*** **Pomel**
SYNONYMS ***Orobanche curvata*** **Pomel;** ***Orobanche icterica*** **Pau;** ***Orobanche elatior*** **subsp.** ***icterica*** **(Pau) A.Pujadas**

Habit: robust, many-flowered cylindrical inflorescence, 9–16(–24) cm, congested above, lax below. **Height**: (20–)25–60(–70) cm. **Stem**: stout, creamy, yellowish, ochre or rosaceous to reddish, glandular-pubescent; scales ovate to triangular, numerous, 15–23 mm long, the uppermost densely hairy-glandular. **Calyx**: segments contiguous, ovate, generally entire, sometimes unequally bidentate, with long acuminate teeth of 4–7 mm; calyx (7–)8–11(–15) mm long. Floral bracts 16–25 mm, similar to the stem bracts, longer than, or sometimes equalling the corolla. **Corolla**: erecto-patent to posteriorly patent, tubular, slightly infundibuliform, dorsal line uniformly curved; (15–)18–22(–25) mm; yellowish, ochre, sometimes pink or more rarely dark red, glandular-pubescent. Upper lip slightly emarginate, sometimes entire, with rounded lobes from porrect to patent; lower lip 3-lobed, with the median lobe somewhat larger than the laterals, oblong, generally deflexed, unequally denticulate, sub-glabrous. **Carpel**: stigma lobes yellow, just touching; style sparsely glandular-pubescent. **Stamens**: filaments inserted 3–4 mm from the base of the corolla, slightly hairy in their lower ½, sub-glabrous or glabrescent in their upper part and with sparse glandular hairs below the anthers; anthers with papillose hairs at the suture. **Host**: exclusively on *Centaurea* spp. (mainly *C. aspera*), occasionally *Rhaponticoides alpina* or *Cheirolophus intybaceus* (Asteraceae). **Ecology and distribution**: dry grasslands, orchards and pine forests. Flowering April–May (June), fruiting May–June. Mediterranean Basin (Algeria, France, Morocco, Spain and Tunisia). **Key diagnostics**: robust, dense rather pale-coloured inflorescence, with tubular corolla with dorsal line uniformly curved; rounded lobes of lower lip; host-specific. **Similar species**: similar and related to *O. elatior*, under which some authors sometimes include the taxon, but *O. leptantha* differs in having shorter stem scales, rounded, converging corolla lower lip lobes, longer floral bracts, lower filament insertion and calyx segments contiguous or rarely connate at the base; limited to the western Mediterranean, parasitising several *Centaurea* species, especially *C. aspera* (not on *C. scabiosa*, the host of *O. elatior*).

Orobanche leptantha
Spain (left, right)

NAME ***Orobanche litorea*** Guss.
SYNONYMS ***Orobanche littoralis*** Lojac.; ***Orobanche minor* var. *litorea*** (Guss.) Bég.; ***Orobanche fuliginosa*** Reuter ex Jord.

Habit: short, stocky spikes; inflorescence ovoid to cylindrical, dense to sub-dense, measuring 3–6(–20) × 2.3–3.5 cm. **Height**: (10–)13–19(–35) cm. **Stem**: swollen at the base, puberulent, reddish-brown, lower stem scales reflexed. **Calyx**: bifid, unequal, just exceeding the corolla; segments subulate acuminate to lanceolate with sub-filiform apex. Floral bracts shorter than the corolla (although longer in bud); violet-brown. **Corolla**: sub-erect with a rather straight dorsal line, sub-glabrous or with minutely pubescent glandular hairs, 0.1–0.2 mm on dorsal surface, and glandular and eglandular hairs, 0.1–0.2 mm on the basal surface; tube 1.1–1.7 cm, white to cream, tinted violet dorsally, with darker veins; upper lip emarginate; lower lip with central lobe larger than the laterals, often with 2 yellow bosses before the throat. **Carpel**: stigma lobes touching, dark purple; style glabrous or sub-glabrous. **Stamens**: geniculate, adaxial filaments inserted 0.3–0.45 cm above corolla base; abaxial filaments inserted 0.15–0.25 cm above the corolla base; anthers violet, (0.1–) 0.15–0.17 cm. **Host**: *Anthemis maritima* (Asteraceae) in the east of the range; *Plantago coronopus* (Plantaginaceae) in the west; possibly other coastal perennials. **Ecology and distribution**: a mainly central Mediterranean species from southern France to southern Greece; plants from south-west Portugal appear to correspond with this species, so possibly more widespread than considered previously. **Key diagnostics**: coastal ecology, short, stocky spikes; floral bracts shorter than corolla; lower corolla lip with yellow bosses; filaments inserted at the base of the corolla tube. **Similar species**: *Orobanche iammonensis* is a similar, smaller plant, always yellow throughout, with a more bulbous (rather than elongated) lower stem, and different host range and distribution (a narrow endemic of Menorca). Coastal forms of *O. minor* can be similar; *O. minor* subsp. *maritima* in particular is similar and closely related, hinting at a widespread coastal lineage made up of locally host-specific entities; further work is required to tease apart these taxa. Populations of an unexamined entity in the Subsection *Minores* in Madeira that grow on *Andryala glandulosa*, which is endemic to Madeira, Porto Santo and the Desertas, are similar. These plants have puberulent, yellowish stems, rather flat-backed corollas, long, apically filiform calyx lobes, and pinkish stigma lobes, and appear to show a superficial similarity to *O. litorea*; these deserve further attention.

Orobanche litorea
Istria, Croatia (left)
photo: Uwe Grabner
Sicily, Italy (right)
Puglia, Italy (page 400)
photo: Uwe Grabner
Portugal (page 401)

NAME ***Orobanche loscosii*** Carlón, M.Laínz, Moreno Mor. & Ó.Sánchez

Habit: robust spikes with flowers all rather congested above; inflorescence (6–)10–23(–31) cm long, shorter than its stem. **Height**: (17–)26–47(–52) cm. **Stem**: glandular-pubescent, yellowish cream with variable brownish-purple tinge; scales (10–)12–22(–24) cm × 4–6 mm, triangular-lanceolate. **Calyx**: bifid, (9–)11–16(–18) mm; distal ⅔ pink-purple with dense yellow glandular hairs; segments unequal, lanceolate with distinctly dark central nerve. Floral bracts (13–)14–20(–25) mm × 4–5 mm, lanceolate, about equalling the corolla. **Corolla**: erecto-patent, campanulate or widely tubular, the outer surface creamy yellow to dull pink, densely glandular-pubescent, purplish venation often present; (17–)18–24 mm. Upper lip bilobate, shallowly divided with lobes erect to recurved with margins undulate, crenate; lower lip lobes subequal, highly divergent and deflexed, the margins undulate, crenate. **Carpel**: stigma lobes dull orange, olive brown to purplish, somewhat divergent; style purplish, glabrous, with occasional glandular hairs and yellow colouration at base. **Stamens**: filaments densely non-glandular-pubescent towards the base, inserted (1–)2–3(–4) mm from base of corolla; anthers cream, ± glabrous. **Hosts**: only on *Echinops ritro* subsp. *ritro* (Asteraceae). **Ecology and distribution**: endemic to the Ebro basin of north-east Spain in grassy scrub. **Key diagnostics**: olive brown stigma, very glandular-hairy, broad corolla and a low filament insertion point. **Similar species**: *Orobanche elatior* and *O. leptantha* have a yellow stigma and more strongly patent corolla angle, and a higher stamens insertion point; *O. centaurina* has a yellow stigma and a straight section to the dorsal line of the corolla; all have different host species, however; *O. centaurina* is also occasionally found on *Echinops*.

Orobanche loscosii
Spain (right)
photo: Gonzalo Moreno Moral

NAME ***Orobanche lucorum*** A.Braun ex F.W.Schultz
SYNONYMS ***Orobanche lucorum* var. *berberidis*** A.Braun ex W.D.J.Koch; ***Orobanche berberidis*** Facchini ex Reut.; ***Orobanche rubi*** Vaucher ex Duby

Habit: somewhat robust stems, often extensively patch-forming. **Height**: 30–50 cm. **Stem**: rather thick below, 0.9–1.4 cm in diameter to 1.5–2.0 cm at the base, slightly swollen beneath, short-hairy, with glandular hairs of 0.2–0.4(–0.6) mm, with yellow glands; ferruginous when dry. **Calyx**: segments simple, subentire or very unequally bidentate, lanceolate, long acuminate, ovate base, with glandular hairs up to 0.3(–0.4) mm; 0.9–1.7 cm long. Floral bracts (1.1–)1.4–2.7 × 0.5–0.7 cm, longer or shorter than the corolla, lanceolate, yellowish, soon turning brown. **Corolla**: tubular, narrowing below the point of insertion of the filaments, erect, sometimes erecto-patent, back uniformly curved, with glandular hairs up to 0.2 mm; 2.0–3.0 cm; tube rather intense yellow or orange-yellow, tinted wine-red, veins sparsely. Upper lip with porrect to patent (not erect or recurved) lobes, bilobed or indistinctly bilobed with ciliate margins with glandular hairs up to 0.2 mm; lower lip with the middle lobe larger than the laterals; lobes broadly obovate. **Carpel**: stigma yellow, later red or purple; lobes just touching; style yellow, with scattered hairs (ca. 0.2 mm) abaxially. **Stamens**: filaments inserted near the base of the corolla, sparsely dilated, and densely hairy at the base. Filaments 1.3–1.6 cm. Adaxial filaments inserted at 1.5–4.5 mm above the base of the corolla, the abaxial ones at 1.5–3 mm; anthers 2.4–2.8 mm. **Host**: *Berberis vulgaris* and other cultivated taxa of this genus (Berberidaceae), need confirmation reports on *Rubus* and *Crataegus* (Rosaceae). **Ecology and distribution**: mountains of central-southern Europe, Alps (France, Austria, Liechtenstein, Switzerland, Germany, Italy and Slovenia). Reports from the Carpathians (Romania) need confirmation; introduced, mainly in gardens, in northern and central Europe (e.g. Denmark, Poland, Czech Republic, Great Britain, Sweden, Finland, Norway and Lithuania). Forest clearings and thickets, on cool damp substrates in shady places, often near water. Flowering June–August. **Key diagnostics**: a yellow to reddish-orange plant, sometimes producing many spikes beneath its shrubby hosts, among ornamental plantings. **Similar species**: *Orobanche lucorum* is superficially similar to *O. caryophyllacea* and *O. rapum-genistae* but lacks the galeate upper corolla lip of these species. Moreover, the flowers are unscented and have yellow stigmas unlike the former, and have stamens inserted higher than the latter. In general appearance, *O. lucorum* is closest to *O. elatior* although that tends to have stouter stems with more densely packed flowers and calyx segments connate below. Common yellow forms of *O. minor* and *O. hederae* have smaller, narrower flowers. *Orobanche lucorum* bears a certain resemblance to *O. haenseleri* but this species has a reddish-orange to pale salmon-coloured corolla with conspicuous veins. *Orobanche flava* has an erect to recurved upper lip and filaments inserted 4–6 mm above the corolla base. Populations of *O. mayeri* in the mountains of the Western Carpathians are also somewhat similar but have smaller flowers with a different shape, and differ in host range.

Orobanche lucorum
UK, in cultivation (left)
Finland (right) photo: Jani Järvi

NAME ***Orobanche lutea*** Baumg.
SYNONYMS ***Orobanche fragrantissima*** Bertol.; ***Orobanche medicaginis*** Duby; ***Orobanche rubens*** Wallr.

Habit: robust; inflorescence usually dense and many-flowered, cylindrical, usually shorter than its stem. **Height**: 20–60 cm. **Stem**: stout (sometimes slender), erect, yellowish-brown, reddish, glandular-pubescent; scales lanceolate, 15–25 mm. **Calyx**: tubular-campanulate, segments equally or unequally bidentate, usually about ⅓ the length of the corolla, 7–12(–15) mm. Floral bracts about as long as or shorter than the corolla, 12–18 mm, broadly lanceolate, usually brownish, glandular-pubescent. **Corolla**: narrowly infundibuliform or infundibuliform-campanulate, erect, with long tube; dorsal line curved from the base, straight in the middle and deflexed near the upper lip, sometimes strongly curved, almost horizontal; 20–32 mm; brownish-yellow, orange, yellow or reddish. Upper lip emarginated or bilobate, keeled, galeate, with erect or porrect lobes; lower lip not ciliate, deflexed and with 3 crenate, rounded lobes. **Carpel**: stigma lobes yellow, rarely orange, reddish, just touching; style glandular-pubescent or sub-glabrous. **Stamens**: filaments inserted 3–5(–6) mm from corolla base, densely pubescent at the base, almost glabrous in the middle, sparsely glandular-pubescent below the anthers; anthers pubescent. **Host**: exclusively on Fabaceae, mainly *Medicago,* especially *M. sativa*, *M. falcata*, and *M.* × *varia*. **Ecology and distribution**: xerothermic grasslands, steppe, rocky, loamy and grassy slopes, meadows, alfalfa plantations, field edges, ruderal habitats, and forest edges. Flowering (April) May–June (July), fruiting (June) July–August. Western Europe (France) to eastern Europe through central and southern Europe, as well as the Caucasus; less abundant in Asia Minor and central Asia. **Key diagnostics**: stout, dense, yellowish-brown to pinkish inflorescence; corolla usually infundibuliform-campanulate, erect, rarely almost horizontally inclined to axis, with characteristic long, relatively narrow tube and galeate upper lip, yellow stigma; host-specific. **Similar species**: an early-flowering species similar to *O. caryophyllacea*, which is differentiated by its broadly campanulate, fragrant flowers with dark purple contrasting stigma, and different host.

Orobanche lutea
Poland (left, right, page 408, 409)

NAME ***Orobanche lycoctoni*** Rhiner

SYNONYMS ***Orobanche flava* var. *albicans*** Rhiner; ***Orobanche lucorum* f. *lycoctoni*** (Rhiner) Rhiner; ***Orobanche aconiti-lycoctoni*** Moreno Mor., G.Gómez, Ó.Sánchez, Carlón & M.Laínz

Habit: usually robust; inflorescence 6.5–12.5 × 2.9–3.2(–3.5) cm; (5–)8–20-flowered, usually dense. **Height**: 24–32(–40) cm. **Stem**: pale yellow, glandular-pubescent; scales 18–20 × 4–7 mm, triangular-lanceolate, glandular-hairy. **Calyx**: (10.4–)14.5–16.8 × (3.5–)4.5–7.8 mm; calyx segments free, imbricate at the base, entire or unequally 2-toothed; teeth lanceolate, acute. Floral bracts 20–25 × 6–9 mm, usually longer than the corolla; yellow-brownish. **Corolla**: tubular-campanulate, almost erect or erecto-patent, strongly constricted below the insertion of the filaments; 18–25 × 6.4–9 mm; pale to very pale, yellow or whitish. Corolla dorsal line strongly arched (but ± straight distally). Upper lip distinctly and broadly 2-lobed, porrect at first, later ± recurved; lower lip almost equally 3-lobed; lobes ± rectangular to spathulate with margin irregularly dentate and not ciliate. **Carpel**: stigma bilobed, with a flattened distal surface, disciform to ± oblong; bright yellow; style glabrous or almost. **Stamens**: filaments inserted 2.5–3 mm abaxial to 3–4 mm adaxial, their whole length densely long-hairy; anthers 2–2.3 × 1.1–1.3 mm; light brown to brown, the upper suture (ca. ½) pilose, distinctly mucronate (white mucro). **Host**: exclusively on *Aconitum lycoctonum* (Ranunculaceae). **Ecology and distribution**: montane and subalpine habitats to the lower alpine belt in deciduous and mixed forests, on rocky and grassy slopes, and in tall vegetation in south-west to central Europe: Cantabrian Mts (Spain), Alps (Austria, France, Germany, Italy, Slovenia and Switzerland). Flowering June–August, depending on elevation. **Key diagnostics**: overall pale yellow-whitish colour, usually robust with a dense inflorescence; bracts usually longer than the corolla; corolla widely tubular-campanulate and strongly arched; ecologically distinct and host-specific. **Similar species**: *Orobanche krylowii* has a slenderer habit, sparser inflorescence, tubular corolla with dorsal line gently and regularly curved; bracts shorter than or equal to the corolla; corolla lobes with finely eroded-denticulate margins; specific to *Thalictrum minus,* occurring from eastern Europe (Russia) to Asia and occasionally in south-east Europe (Albania).

Orobanche lycoctoni
Spain (left, right)
photos: Gonzalo Moreno Moral

NAME ***Orobanche mayeri*** (Suess. & Ronniger) Bertsch & F.Bertsch

SYNONYMS ***Orobanche alsatica* var. *mayeri*** Suess. & Ronniger; ***Orobanche alsatica* subsp. *mayeri*** (Suess. & Ronniger) Kreutz

Habit: usually stout; inflorescence cylindrical, usually dense and many-flowered. **Height**: 35–45(–60) cm. **Stem**: slender to stout, erect, yellow, dark brown, pinkish-brown, purple, glandular-pubescent; scales (15–)18–20(–22) mm, triangular below, above lanceolate. **Calyx**: segments usually regularly bifid, about ⅓ to ½ the size of the corolla, glandular, the front halves of the calyx usually free; (9–)10–14(–15) mm. Floral bracts ca. ⅔ of the length of the corolla, lanceolate, glandular-pubescent. **Corolla**: erect to erecto-patent, usually tubular, with funnel-shaped throat; dorsal line evenly curved throughout and almost horizontal at the level of upper lip; (12–)15–17(–19) mm; bright to dull yellow, pinkish to yellowish-brown, or pink-brown. Upper lip usually slightly emarginate with broad, erect lobes; lower lip 3-lobed, irregularly dentate and deflexed. **Carpel**: stigma lobes yellow or orange, touching; style almost glabrous below and sparsely glandular above. **Stamens**: filaments inserted (2–)3–4 mm above corolla tube, glabrous below and sparsely glandular above near anthers; anthers pubescent at the line of fusion. **Host**: Apiaceae: *Laserpitium latifolium* (*Pimpinella major* subsp. *rhodochlamys* needs confirmation). **Ecology and distribution**: mountain meadow, rocky ledges, relict pine groves, thinned spruce, oak-hornbeam forests, and relict montane xerothermic grasslands on dolomites and limestone, 700–1,000(–1,500) m. Flowering end June–end July (August). Known only from single localities in Germany, Slovakia, and Poland (needs confirmation in Romania); a rare species associated with the Western Carpathian flora. **Key diagnostics**: a stout, tall, yellow, pinkish to yellowish-brown plant with a cylindrical, dense and many-flowered inflorescence; flowers rather small relative to the plant, erect to erecto-patent; corolla tubular with dorsal line evenly curved throughout; distinct mountain ecology and host. **Similar species**: formerly treated as variety or subspecies of *Orobanche alsatica* but found to be closer to *O. flava* and *O. lucorum*. *Orobanche mayeri* differs from *O. alsatica* in the frontally free calyx segments, smaller, erect to erecto-patent flowers, corolla morphology, almost glabrous style, lower filament insertion point, host and ecology.

Orobanche mayeri
Poland (left, right)

NAME ***Orobanche minor*** Sm.
SYNONYMS ***Orobanche apiculata*** Wallr.; ***Orobanche concolor*** Duby; ***Orobanche trifolii-pratensis*** F.W.Schultz

Habit: slender; inflorescence usually lax and interrupted below, dense above. **Height**: (5–)10–70 cm tall. **Stem**: slender, sometimes stout below, red-brown, violet-purple, brown-red or tinged with red (rarely all yellow), densely glandular-pubescent; scales few, lax, reflexed. **Calyx**: equally to unequally bifid, rarely entire, 6–14 mm. Floral bracts 6–22 mm, ovate-lanceolate to lanceolate-subulate, acuminate, equal to or just exceeding the corolla. **Corolla**: curved regularly throughout (or sharply inflected at the base then curved), cream, purplish suffused to a variable extent, hairless to glandular-hairy; 10–18(–20) mm; upper lip emarginate or slightly bilobed with the lobes directed forwards; lobes of lower lip subequal or with the central the largest, all lobes crisped and denticulate, without prominent bosses. **Carpel**: stigma lobes typically pinkish-purple, to reddish-purple (or yellow in all-yellow variants), just touching at the base to partially united; style sparsely glandular-pubescent or glabrous. **Stamens**: filaments inserted 2–3.5 mm above the base of the corolla tube, sparsely pubescent at the base, sparsely glandular-pubescent beneath the anthers; anthers often pubescent at the line of fusion. **Host**: on numerous herbs (rarely shrubs) from many families, especially Fabaceae and Asteraceae. Common hosts include *Trifolium* species (*T. pratense, T. medium* and *T. arvense*), *Lotus corniculatus* and *Medicago sativa*; also on cultivated hosts. **Ecology and distribution**: very widespread and common: native to Eurasia, now cosmopolitan in temperate biomes; occurs in disturbed habitats, gardens, roadsides and fixed dunes, as well as mountain pastures. **Key diagnostics**: a highly variable species, with many forms, which can make identification difficult. Typically purplish, rather small-flowered and slender; corolla tubular, not widened upwards, its back curved ± evenly throughout. **Similar species**: this is the common species in the complex Subsection *Minores.* The features emphasised above and mostly ruderal ecology are key. *Orobanche picridis* is similar and can share ruderal habitats; it is a paler plant with slender stems, darker stigma, straighter-backed corolla and very slender bracts and calyx lobes. Pale forms of *O. minor* with attenuated calyx lobes and rather hairy filaments can be confusing, but these lack the purple to blackish stigma lobes. *Orobanche hederae* can also overlap in ecology and distribution and differs in its yellow stigma, corollas with a distinct constriction behind the corolla mouth and more acute corolla lower lip lobes. Infraspecific taxa can be very difficult to distinguish; host and ecology are important considerations.

Orobanche minor var. *heliophila*
UK (left)
Orobanche minor subsp. *minor*
UK (right)

Infraspecific taxa (selected)

Subsp. *minor* is the widespread form of the species and has numerous, often host-defined forms: **f. *lutea*** is the common all-yellow form; coastal morphs of this in the UK are sometimes stunted, with globose heads. **Var. *compositarum*** has corolla conspicuously sub-erect (especially so in fruit), slender, 3.5–5 mm in diameter, pale and sub-glabrous. Hosts Asteraceae (especially *Crepis*), rarely other families; known only from the UK but possibly elsewhere. **Var. *heliophila*** is specific to *Brachyglottis* in landscaped environments; clumped, perennating; corolla basally white, strongly flushed purple distally; filaments densely hairy at the base. Reported from the UK, possibly elsewhere in northern Europe. **Var. *pseudoamethystea*** is specific to *Eryngium maritimum* and other halophytes, exclusively in maritime habitats. Spike often dense; corolla (>15 mm) strongly tinted purple, with flattish dorsal line, geniculate at the tip, bilobed upper lip and a high filament insertion (≥3 mm above the base). Known locally from south British and northern French coasts.

Subsp. *maritima* is glandular-puberulent, short (to 35 cm). Bracts not exceeding the corolla; calyx short, 6–10 mm, often entire. Lower lip with middle lobe largest, reniform, with 2 prominent yellow bosses. Stigma lobes partially fused. Host *Daucus carota* subsp. *gummifer* (rarely *Plantago* or Asteraceae); known from south British and western French coasts. Resembles *O. litorea*. **Var. *hypochoeridis***, a yellow plant, has a dense often sub-globose spike, corollas <12 mm, calyx short, usually entire. Known only from the Channel Islands on Asteraceae tribe Cichorieae, possibly extinct.

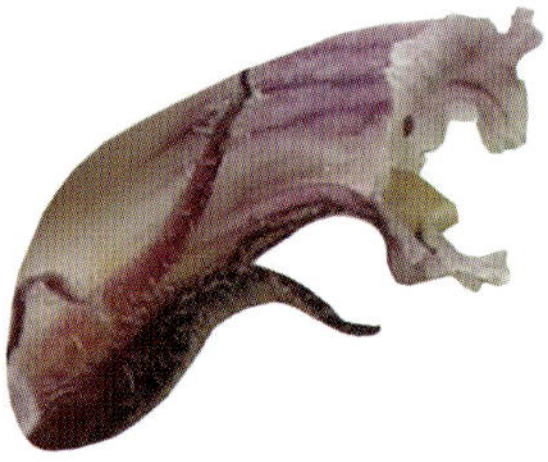

O. minor subsp. *maritima*

O. minor subsp. *minor* var. *heliophila*

O. minor subsp. *minor* var. *pseudoamethystea*

Orobanche minor var. *pseudoamethystea* UK (right) photo: Rebecca Wheeler
Orobanche minor subsp. *maritima* UK (page 418)
Orobanche minor f. *lutea* (globose form) UK (page 419)
Orobanche minor var. *compositarum* UK (page 420)
Orobanche minor Sardinia, Italy (page 421)

NAME ***Orobanche mlokosiewiczii*** Piwow., Ó.Sánchez & Moreno Mor.

Habit: usually slender; inflorescence (3.0–)7.5–9.0(–15) cm, cylindrical to slightly ovoid, shorter than its stem, ± dense, sometimes lax. **Height**: (16–)30–40(–53) cm. **Stem**: slender, pale white-yellow or pinkish-brown pale, densely glandular-pubescent above; scales ovate-lanceolate (triangular) to lanceolate and narrowly lanceolate, (8–)12–17(–21) mm. **Calyx**: segments, free, clearly separate, simple, entire, narrow, with teeth long, acuminate almost filiform at apex and with bases rather ovate; (7–)9–10(–13) mm long, shorter than ½ the width of the corolla tube. Floral bracts (13–)16–17(–22) mm, shorter than or equal to the corolla, narrowly lanceolate, with dense white or pale yellow glandular hairs. **Corolla**: erecto-patent, campanulate, rarely tubular-campanulate, with a broad, open throat, usually strongly inflated above the insertion of the filaments, and strongly constricted below; the dorsal line slightly curved, in the upper part straight and strongly patent; (17–)20–23(–25) mm long, (7–)10–12(–14) mm in diameter in the central part; pale white-yellow, whitish, rarely whitish-pink, yellowish-white, with dirty white, pale pink or pale brown veins; sensitive to damage and browning; externally glandular-pubescent with white or pale yellow glandular hairs. Upper lip slightly emarginate, later distinctly upward-curved and patent, rarely with 2 broad lobes, emarginated, sparsely glandular-pubescent also in the inner part; lower lip with 3 ovate, or ± rectangular to ± triangular lobes, central lobe larger than laterals, irregularly dentate on margins, sparsely glandular-pubescent also in the inner part. **Carpel**: stigma lobes elongated spherical to ovate, bright yellow, touching; style with sparse (more abundant on the upper part) short glandular hairs. **Stamens**: filaments obliquely inserted 3.5–5 mm above the corolla base, densely villous in the base to ± middle part, not glandular, upper part with dense short, usually glandular hairs; anthers distinctly hairy along ¾ of suture. **Host**: *Aconitum* spp. (Ranunculaceae), especially *A. cymbulatum*. Possibly also on *Senecio propinquus* (Asteraceae) in Georgia. **Ecology and distribution**: thermohygrophilous, subalpine, tall, herbaceous vegetation, near waterfalls or on rocky slopes. Flowering July–August, fruiting August. Probably Caucasian endemic, in the Greater Caucasus: Georgia and Russia (Karachay-Cherkessia). **Key diagnostics**: campanulate, pale flowers with a conspicuously open throat and tube narrowed sharply to the base; lacking distinct upper lip lobes; host-specific. **Similar species**: *Orobanche lycoctoni* and *O. krylowii* also grow on Ranunculaceae but differ in having free calyx segments, smaller bracts, different corolla dorsal line, and higher filament insertion point. *Orobanche cicerbitae* differs in having a tubular and strongly curved corolla, distinct lobe morphology, and specificity for Asteraceae.

Orobanche mlokosiewiczii
Georgia (left, right)

NAME ***Orobanche montserratii*** A.Pujadas & D.Gómez

Habit: robust, erect stems with short, compact inflorescences (far shorter than their stems) of rather small flowers relative to the plant. **Height**: (25–)30–50 cm. **Stem**: strongly swollen beneath; dull yellowish-brown; scales rather numerous, brown, erect, 18–21 × 4–5 mm. **Calyx**: segments completely free, unequal (rarely entire), ovate at the base, subulate, with a marked central nerve. Floral bracts 18–21 mm, equalling or exceeding the corolla. **Corolla**: tubular, patent with a markedly curved dorsal line; 15–21 mm, upper lip emarginate, porrect; lower lip with ovate lobes, the middle largest; lobes scarcely deflexed and irregularly toothed, ciliate, with glandular hairs along the margins; dull brownish-yellow or pink. **Carpel**: stigma lobes yellow, somewhat divergent. **Stamens**: filaments inserted (2–)2.5–3 mm above the corolla tube base, densely hairy beneath. **Host**: large Apiaceae (especially *Laserpitium,* such as *L. latifolium* and *L. nestleri;* reported also on *Angelica* and *Imperatoria*). **Ecology and distribution**: mountains in France and Spain; dry grasslands and rocky places, usually at 900–1,450 m. Flowering June–August. **Key diagnostics**: rather dull beige-coloured congested spikes of small flowers relative to the plant; free calyx segments; floral bracts equalling or exceeding the corolla. **Similar species**: most similar to *O. alsatica, O. bartlingii* and *O. mayeri* but distinguished by its very unequal, free calyx segments (not subequal to entire), short, porrect (rather than spreading) upper corolla lip lobes, and specificity for *Laserpitium* (also a host of *O. mayeri*).

Orobanche montserratii Spain (above, right) photos: Antonio Pujadas Salvá

NAME **Orobanche owerinii** (Beck) Beck
SYNONYM ***Orobanche crenata* var. *owerinii*** Beck

Habit: slender or slightly robust; inflorescence usually short-cylindrical to ovoid, ± lax, usually few-flowered and shorter than or equalling its stem. **Height**: 15–60 cm. **Stem**: usually slender, glandular-pubescent, with mixture of longer, stiff hairs, reddish-brown, rarely yellowish-brown; scales lax, lanceolate, 15–25 mm long. **Calyx**: segments free, entire or less frequently bidentate, narrowly subulate, up to 15 mm, shorter than their corolla tube. Floral bracts lanceolate, equalling or slightly shorter or longer than the corolla, 15–25 mm. **Corolla**: campanulate-infundibuliform; dorsal line almost straight but evenly curved at the proximal part and, sometimes, slightly bent forward distally; throat open; 15–20(–30) mm; pale dull creamy-violet, rosaceus to purple, the interior paler (whitish or yellowish) and veined. Upper lip with 2 short, broad lobes, emarginate, usually erect; lower lip with 3 large suborbicular lobes, central lobe slightly larger, irregularly and strongly dentate, deflexed. **Carpel**: stigma lobes divergent, reddish to deep purple; style sub-glabrous. **Stamens**: filaments inserted 3–4 mm above the corolla base; hairy along ½ (or ⅓) of the base and sparsely glandular-pubescent beneath the anthers; anthers sub-glabrous. **Host**: mainly on *Trifolium*, especially *T. canescens, T. ambiguum* and *T. trichocephalum;* occasionally *Vicia iberica* (Fabaceae) rarely on *Leontodon hispidus* (Asteraceae). **Ecology and distribution**: subalpine and alpine meadows, rocky and grassy slopes, pastures, roadsides, rarely forests, usually (600–)1,800–2,400(–2,600) m. Flowering June–July (August), fruiting August (September). Caucasus (Georgia, Armenia, Russia and Azerbaijan), and probably also in neighbouring Turkey and Iran. **Key diagnostics**: a polymorphic species with variable colouration; inflorescence usually short and few-flowered; corolla campanulate-infundibuliform; preference for subalpine and alpine meadows and specificity for *Trifolium*. **Similar species**: *Orobanche laxissima* differs in having a long cylindrical, rather lax but many-flowered inflorescence and calyx segments entire or bidentate; corolla tubular-infundibuliform, longer, with lower filament insertion point; usually occurs in forests and parasitic on trees. *Orobanche crenata* differs in its broadly campanulate, white corolla with darker veins, and fragrance.

Orobanche owerinii
Georgia (left, right, page 428, 429)

NAME ***Orobanche pancicii*** **Beck**

Habit: slightly robust; inflorescence 5–15 cm, dense, often lax below. **Height**: 15–50 cm. **Stem**: slightly robust, red, pinkish-brownish, yellowish-brown, dark purple, glandular-pubescent; scales 13–25 mm, lanceolate, dense at base, sparse above. **Calyx**: 12–15 mm, segments unequally bidentate to entire, 5–8-veined. Floral bracts ovate and long-acuminate to ovate-lanceolate, equalling or shorter than the flowers. **Corolla**: (15–)20–23 × 6–9 mm, erect, tube broadly tubular-campanulate, evenly curved or almost straight in lower ⅔; pale yellow with purplish veins, or dark purple externally and brighter (yellowish or pale pink) internally, densely covered with short, dark purple glandular hairs. Upper lip slightly emarginate; lower lip lobes crenate to almost entire, glandular-ciliate. **Carpel**: stigma yellow, rarely pink or purple; style sparsely glandular-pubescent, especially below the stigma, purple. **Stamens**: filaments hairy at the very base; anthers hairy at the line of fusion. **Host**: Caprifoliaceae, mainly *Scabiosa* and *Knautia*; reports on other families require investigation. **Ecology and distribution**: pastures, meadows, subalpine grassland, tall herbaceous vegetation, forests and forest margins, usually at the subalpine belt, ca. 1,600–1,900(–2,200) m. Known to occur from Austria to the Balkan Peninsula and Hungary. Flowering (June) July–August. **Key diagnostics**: corolla densely covered with short, dark purple glandular hairs; outer and inner parts of the corolla with differing colouration; flowers with a distinct smell of carnations; montane ecology and host-specific. **Similar species**: *Orobanche reticulata* differs mainly in its corolla shape, purple to dark violet stigma, lack of fragrance, and host range.

Orobanche pancicii Hungary (above) photo: David Schmidt, Austria (right) photo: Stefan Mayr

NAME ***Orobanche picridis*** F.W.Schultz

SYNONYMS ***Orobanche loricata* var. *picridis*** (F.W.Schultz) Bég.; ***Orobanche minor* var. *picridis*** Franchet; ***Orobanche artemisiae-campestris* subsp. *picridis*** (F.W.Schultz) O.Bolòs, Vigo, Masalles & Ninot

Habit: slender, rarely stout, with cylindrical, usually dense, many-flowered inflorescence. **Height**: (10–)25–50(–70) cm. **Stem**: virtually hairless to glandular-hairy, pale yellowish, often tinged dull purple, with few brown scales below, and scarcely swollen at the base. **Calyx**: segments entire or unequally bifid; the segments slender, acuminate, 1-veined or 3-veined, the outer veins very indistinct; 10–15 mm. Floral bracts 12–20 mm, narrowly ovate-acuminate, recurved, equalling or exceeding the flowers, papery dark brown, contrasting with the pale flowers. **Corolla**: the back of the corolla curved at base, then virtually straight; 14–22 mm; very pale, ivory to yellowish, veined purple-lilac distally, virtually hairless to glandular-hairy, the upper lip emarginate to bifid, porrect to erect, ± forming a standard; the lobes of lower lip subequal, or the middle the largest, rather square in shape; all lobes crisped and toothed; corolla with brown papery patches upon opening. **Carpel**: stigma lobes dark purple (to blackish), contrasting with the pale flowers, just touching; style sparsely glandular-pubescent. **Stamens**: inserted (2–)3–5 mm above the base of corolla; filaments with conspicuously dense white hairs below. **Host**: mostly on *Picris hieracioides* (Asteraceae), sometimes on *Daucus carota* (Apiaceae); occasionally on other herbs including *Pilosella officinarum* and *Pentanema squarrosum* (Asteraceae). **Ecology and distribution**: habitats range from sea cliffs and fixed dunes to ruderal environments in the north, and xerothermic grasslands, hill slopes, fallow lands and scrub further south. Flowering rather late, (May) mid-June–end July (August). Widespread, although local across south-west Europe, from Spain, across the Mediterranean Basin, France, southern Germany and Austria, to the Balkans, north to the UK and the Netherlands, Czech Republic, Slovakia, southern Poland, Romania and Ukraine. Reported from Asia Minor, the Transcaucasian region and northern Africa (to be investigated). **Key diagnostics**: a pale plant with contrasting bracts and stigma lobes, slender, with flowers rather large relative to the slender stems; calyx lobes long and slender, the upper segment with one conspicuous vein only; floral bracts ± exceeding the flowers and often conspicuous at the top of the spike. **Similar species**: similar to *O. minor* but with paler corolla (contrasting with the dark stigma) with the upper lip porrect; filaments inserted ≥3 mm above the base of the corolla tube, their bases with a shaggy covering of whitish hair; calyx with long filiform limbs. Much-confused with *O. artemisiae-campestris*, which is a very different plant with a yellowish corolla with reddish colouration, open throat, calyx-teeth divided up to ⅔ length, almost to the base, and specificity for *Artemisia campestris*. The poorly known Turkish endemic *O. hadroantha* may be synonymous. *Orobanche crenata* differs in its broadly campanulate corolla with darker veins, and fragrance.

Orobanche picridis
South Wales, UK (left, page 434, 435)
Puglia, Italy (right)

NAME ***Orobanche pubescens*** d'Urv.
SYNONYMS ***Orobanche minor*** **var.** ***pubescens*** (d'Urv.) Meikle; ***Orobanche versicolor*** F.W.Schultz

Habit: slender; inflorescence usually short, cylindrical to oblong and usually shorter than its stem. **Height**: 10–30(–60) cm. **Stem**: abundantly glandular-pilose or villous, yellowish, pinkish, brownish, reddish; scales 10–25 mm, oblong or linear-lanceolate, glandular-pubescent to villous. **Calyx**: segments free, undivided or bidentate, teeth narrow, long-acuminate or filiform, long glandular-pilose, usually equalling the corolla tube; 8–16 mm. Floral bracts equalling corolla or longer, usually white glandular-villous, 10–20 mm long, lanceolate, acuminate. **Corolla**: erecto-patent; dorsal line nearly straight from a curved base, narrowly tubular to tubular-campanulate, ± uniformly curved, at first erect-spreading, then stretched or curved forward; 10–20 mm; yellowish-white, pinkish-white, violet-white or violet-brown towards the limb, long lanate-pilose outside, especially in the upper lip. Upper lip ± entire, usually curved downwards; lower lip 3-lobed, the middle lobe largest, all lobes glabrous or sparsely ciliate along the margins. **Carpel**: stigma lilac, pink, violet, purple, rarely blue or yellow; style sub-glabrous. **Stamens**: filaments inserted 3–4 mm above corolla base; long-pilose in lower part, almost villous, usually glabrous above; anthers glabrous or sub-glabrous. **Host**: many species from various families, often Asteraceae (e.g. *Galactites tomentosus*, *Hyoseris radiata*) and Apiaceae (e.g. *Orlaya*, *Caucalis*), probably also Fabaceae; detailed host range requires verification. **Ecology and distribution**: xerothermic grasslands and shrubs, rocky slopes, ruderal habitats. Flowering (end April) May–June. Eastern Mediterranean, Balkans, Crimea, Caucasus to Asia Minor. **Key diagnostics**: abundantly glandular-pilose or villous, with admixture of long white hairs. **Similar species**: superficially similar to *O. minor* which overlaps in host and ecology but lacks the conspicuous indumentum of *O. pubescens*.

Orobanche pubescens
Crete, Greece (left, right, page 438, 439)

NAME ***Orobanche raddeana*** Beck

SYNONYMS ***Orobanche alba* var. *raddeana*** (Beck) Beck; ***Orobanche campanulae*** Rätzel, Ristow & Uhlich

Habit: usually slender; inflorescence short, cylindrical-ovoid or capitate, usually lax and relatively few-flowered, shorter than its stem. **Height**: 15–30 cm. **Stem**: slender, yellowish-brown, often with a reddish tinge, glandular-pubescent, slightly thickened at the base; scales up to 20 mm long, broadly lanceolate, usually dense and numerous. **Calyx**: segments usually entire, free, from an ovate base, narrowly lanceolate, slightly falcate, shorter or as long as ½ the length of the corolla; 8–16 mm. Floral bracts ovate-lanceolate, with narrowly lanceolate tip, up to 20 mm long. **Corolla**: erecto-patent, tubular to tubular-campanulate, dorsal line ± regularly curved in the middle portion, almost equally concavely arcuate; 16–23 mm; dark-red to reddish, pinkish, rarely creamy-pinkish-violet, with dark veins, scattered with pale glandular hairs. Upper lip shallowly 2-lobed to emarginate, lobes broadly roundish, projected forward or erect (rarely reflexed), irregularly dentate-lobate and glandular-pubescent; lower lip 3-lobed, lobes roundish, ± the same size, darker veined, marginally obtusely irregularly dentate-lobate, glandular-ciliate. **Carpel**: stigma lobes purple or dark-red; style densely glandular-pubescent. **Stamens**: filaments inserted 2–3 mm above base of corolla tube, densely hairy at the base and usually along the entire length; anthers hairy or sub-glabrous. **Host**: exclusively on Campanulaceae (e.g. *Campanula* and *Asyneuma*), especially *C. alliariifolia, C. collina*, *C. tridentata* and *A. campanuloides*. **Ecology and distribution**: upper montane, subalpine to alpine habitats, especially in the Greater Caucasus, rarely in the Lesser Caucasus: rocky and stony slopes, meadows, pastures, rocky ledges, forest edges and scrub, usually (800–)2,000–2,500(–2,700) m. Flowering (June) July–August, fruiting August (September). Endemic to the Caucasus: Russia, Georgia and Azerbaijan. Needs confirmation in other Caucasian countries and surrounding areas such as Turkey and Iran. **Key diagnostics**: restricted to Campanulaceae in subalpine to alpine habitats; short or capitate usually lax and relatively few-flowered inflorescence; corolla reddish, tubular to tubular-campanulate with dorsal line ± regularly curved at the middle part. **Similar species**: the most similar species is *O. alba*, which differs in its broadly campanulate corolla with scattered, ± sessile dark glands (giving a punctuated appearance), dorsal line ± almost straight in the middle portion, its stamens ± (not densely) hairy at the base, calyx segments often subulate at the tip, purple (or yellow/orange) stigma and preference for Lamiaceae. Recently, the name *O. campanulae* Rätzel et al., was proposed for this taxon, however here we retain *O. raddeana* as the valid name.

Orobanche raddeana
Georgia (right, page 442, 443)

NAME ***Orobanche rapum-genistae*** Thuill.
SYNONYMS ***Orobanche cytisi-scopari*** Vaucher ex Duby; ***Orobanche rapum*** Wallr.

Habit: robust, usually clumped; inflorescence cylindrical, many-flowered, dense above and lax below. **Height**: (15–)25–70(–90) cm. **Stem**: always stout and much-swollen beneath, erect, yellow, yellowish-brown, orange-brown, brownish, pinkish-brown, reddish-brown, glandular-pubescent; scales triangular to lanceolate. **Calyx**: usually bifid with free, unequally bidentate (rarely entire) segments with elongated acute tips, glandular-pubescent, shorter than or ½ as long as the corolla, 8–15 mm. Floral bracts 15–30 mm, lanceolate, acute, usually much longer than the corolla, sparsely glandular-pubescent. **Corolla**: sub-erect to erecto-patent, broadly tubular to usually campanulate; dorsal line evenly curved over its entire length and helmet-shaped near the top; 20–25 mm; yellowish, reddish-brown, violet-brown, pinkish-brown, dark-veined, usually glandular-pubescent. Upper lip almost entire, helmet-shaped, lobes porrect or slightly raised; lower lip lobes almost equal or middle lobe slightly largest, rounded, crenate and deflexed. **Carpel**: stigma lobes yellow, markedly distant; style usually glandular-pubescent along its entire length, with longer glandular hairs below the stigma. **Stamens**: filaments inserted up to 2 mm above the base of the corolla tube, glabrous at the base, glandular-pubescent below anthers; anthers glabrous or almost so. **Host**: always on woody Fabaceae, especially *Cytisus scoparius*, rarely *C. striatus*, *C. multiflorus* or other *Cytisus* species, *Genista* and *Ulex*. **Ecology and distribution**: xerothermic grasslands, sandy, coastal scrub, cliffs, open forests. Early flowering, usually May–June (July). Very widespread across western Europe south from the UK to central Germany to the Mediterranean coasts, through the mountains of Sicily to north-west Africa. **Key diagnostics**: easily identified by its stature and host (shrubby Fabaceae). Key features include the almost entire upper corolla lip, stamens with hairless filament bases, inserted virtually at the base of the corolla tube, and well-separated yellow stigma lobes. **Similar species**: polymorphous species with many forms. *Orobanche rigens* (also considered conspecific) has flowers almost to ground-level, corolla glabrous or sub-glabrous, broad corolla, gradually enlarged towards the throat, filaments and style glabrous, and is restricted to Sardinia and Corsica in association with endemic hosts such as *Genista corsica*. *Orobanche elatior* is also robust and brownish (typically tinted orange) but has a finely toothed upper corolla lip (not virtually entire), and stamens inserted well above the base of the corolla tube, with hairy filament bases. *Orobanche caryophyllacea* has a somewhat similar corolla shape and may have a similar colour but has a distinct scent and purplish stigma lobes, as well as a different host.

Infraspecific taxa (selected)

Subsp. *rapum-genistae* is the widespread, common form of the species, which has a characteristic almost entire upper corolla lip.

Subsp. *benthamii* from south-west Europe and northern Italy is distinguished by its bracts at least ½ the length of the corolla and the corolla ± erect with a 2-lobed upper lip; generally regarded as not worthy of subspecific recognition.

Orobanche rapum-genistae
UK (right, page 446)
Spain (page 447)
UK (pages 448–449) photo: Grahame Preston

NAME ***Orobanche reticulata*** Wallr.
SYNONYMS ***Orobanche cirsii*** Fr.; ***O. pallidiflora*** Wimm. & Grab.

Habit: usually robust; inflorescence cylindrical or short, from few flowers to many-flowered and dense, laxer below. **Height**: (15–)30–70(–90) cm. **Stem**: slender or robust, erect, brown, yellowish-brown, violet-brown, glandular-pubescent; scales triangular to lanceolate, (10–)15–25 mm. **Calyx**: segments usually entire, rarely unequally bidentate, ovate to lanceolate, about ½ as long as the corolla, brown, red-brown or violet, contrasting with the corolla; 7–12 mm. Floral bracts 12–20 mm, about as long as or longer than the corolla, lanceolate, usually dark brown, reddish-brown, glandular-pubescent. **Corolla**: erecto-patent, later almost horizontal, broadly tubular to tubular-campanulate; tube somewhat broadened above the insertion of the stamens, the back curved strongly above the base and behind the upper lip; 14–20(–25) mm; violet, variable shades of cream tinted violet, with numerous dark violet or yellow glands; often paler beneath. Upper lip lobes broad, rounded and erecto-patent, helmet-shaped, veined; lower lip lobes deflexed, with rounded, usually sharply plicate, crenate lobes with violet veins, the middle lobe a little longer than the laterals. **Carpel**: stigma lobes red-violet, purple or orange-brown, just touching at the base; style sparsely glandular-pubescent or glabrous. **Stamens**: filaments inserted 2–4 mm above the base of the corolla tube, glabrous or sparsely pubescent; anthers sparsely glandular-pubescent at the line of fusion. **Host**: usually on *Carduus* and *Cirsium* (Asteraceae), especially *Carduus acanthoides*, *C. crispus*, *C. defloratus*, *Cirsium oleraceum*, *C. palustre*, *C. erisithales* and *C. arvense*; reports of parasitism on Caprifoliaceae, e.g. *Scabiosa*, require confirmation. **Ecology and distribution**: found in a range of habitats, from plains to subalpine and alpine areas; on rocky grasslands, in humid meadows and herb thickets, ruderal habitats and field edges, usually on alkaline soil. Flowering rather late, June–July, sometimes in higher altitudes to August. Widespread: throughout Europe, north to the UK, east to Turkey; at higher elevations in the south. **Key diagnostics**: rather robust, creamy or purple spikes with flowers with entire calyx segments with dark colour contrasting with corolla tube, dark glands and dark purple stigma and upper lobes; corolla is curved strongly behind the upper lip; preference for thistles. **Similar species**: mountain plants sometimes separated as *O. reticulata* s.str. (smaller in size, flowers more densely dark-glandular, filaments more densely glandular above, and a narrower host range but including examples of the Caprifoliaceae), while those at lower altitudes are referred to as *O. pallidiflora* (taller, lighter-coloured plants). Generally these are considered ecotypes of the same species. Robust specimens of *O. minor* can resemble *O. reticulata* (and also parasitise thistles), but the former have narrower, typically less arched corolla tubes that lack dark glands, and the filaments are never glandular-pubescent above.

Orobanche reticulata
Czech Republic (left) photo: František Lamla
Poland (right, page 453)
UK (page 452) photo: Grahame Preston

NAME ***Orobanche rigens*** Loisel.
SYNONYMS ***Orobanche rapum-genistae* subsp. *rigens*** (Loisel.) Arcang.; ***Orobanche thyrsoidea*** Moris ex Bertol.

Habit: robust; inflorescence ovoid or cylindrical, the tip rounded and dense. **Height**: 15–38 cm. **Stem**: much-thickened at the base and with densely overlapping bracts; yellow-reddish or red-purple, pubescent; lower scales broadly ovate-rounded, the uppermost oblong, glabrous or sparsely hairy, up to 1.5 cm long, usually dense. **Calyx**: segments anteriorly shortly connate or contiguous from a bidentate ovate base; shorter than ½ the corolla, ± hairy and glandular-hairy; 5–15 mm. Floral bracts glabrous or the uppermost glandular-hairy. **Corolla**: erecto-patent, tubular to campanulate, broad, gradually enlarged towards the throat, ± ventricose above the insertion of the stamens, usually glabrous, sparsely glandular or hairy; the dorsal line curved to the upper lip; (11–)20–27 mm; purplish, reddish-purple, violet, pinkish-brown. Upper lip entire or emarginate, fornicate, carinate, rounded with short reflexed lobes; lower lip equally 3-lobed, lobes elliptical, rarely rounded, glabrous or sub-glabrous with denticulate, glandular-hairy margins. **Carpel**: stigma lobes yellow, lemon, orange, divergent, touching at the base; style usually glabrous. **Stamens**: filaments inserted up to 2 mm above the base of the corolla tube, glabrous; anthers glabrous or almost so. **Host**: woody Fabaceae only: *Genista corsica*, *G. salzmannii* and *G. sulcitana*. **Ecology and distribution**: open oak and birch forests, xerothermic grassland, rocky slopes, coastal scrub, from sea level to montane habitats. Flowering April–July. Endemic to Sardinia and Corsica, also reported from Sicily, possibly in error. **Key diagnostics**: robust, stems with copious scales and flowers to almost ground-level; corolla glabrous or sub-glabrous, broad, gradually enlarged towards the throat; filaments and style glabrous; flowers foetid. Associated with island endemic hosts. **Similar species**: some authors include *O. rigens* in the more widely distributed *O. rapum-genistae*, which differs in its stature, glandular-pubescent corolla, bracts usually much longer than the corolla, and glandular-hairy filaments and style.

Orobanche rigens
Corsica, France (left, right, page 456)
Sardinia, Italy (page 457)

NAME ***Orobanche salviae*** F.W.Schultz ex W.D.J.Koch

Habit: usually stout, rarely slender; inflorescence cylindrical, usually dense, shorter than its stem. **Height**: 10–50 cm. **Stem**: slender, rarely stout, yellowish-brown or brown, glandular-pubescent; scales triangular to lanceolate, dense below, sparser below the inflorescence. **Calyx**: segments entire or unequally bidentate, ca. ½ as long the corolla, reddish. Floral bracts slightly longer than corolla, lanceolate and glandular-pubescent. **Corolla**: tubular, erecto-patent to almost horizontal; 15–23 mm long; slightly inflated above the filament insertion, densely pale glandular-hairy; yellow-brown, pinkish-brown, brown. Dorsal line of the corolla evenly curved along its length (rarely almost straight). Upper lip entire, emarginated or bilobate, with porrect lobes; lower lip 3-lobed; lobes rounded, crenate, middle lobe slightly longer. **Carpel**: stigma yellow or orange, densely glandular-pubescent; style densely glandular-pubescent. **Stamens**: filaments inserted 3–5 mm above corolla tube; densely pubescent at the middle and sparsely glandular-pubescent or glabrous below the anthers; anthers pubescent at the line of fusion. **Host**: *Salvia* spp., usually *S. glutinosa* (Lamiaceae). **Ecology and distribution**: deciduous or mixed mountains forests in the Catalan Pyrenees through the Alps to the Pindos Mountains, from western and especially central Europe to the Balkans. Flowering July–September. **Key diagnostics**: yellow-brown or brown colouration, corolla densely pale glandular-hairy with the dorsal line evenly curved; late flowering; ecologically distinct and host-specific. **Similar species**: somewhat similar to *O. lucorum* and *O. flava* but differs in the features stressed above and in its host.

Orobanche salviae Croatia (above, right) photos: Marko Doboš

NAME ***Orobanche sanguinea*** C.Presl
SYNONYMS ***Orobanche crinita*** Viv.; ***Orobanche loti-cytisoidis*** F.W.Schultz; ***Orobanche nebrodensis*** Tineo

Habit: slender, small-flowered spikes; rarely clumped. **Height**: to 30 cm. **Stem**: rather slender, swollen at the base, glandular-pubescent, light to dark purplish-red; scales 1.2–3.5 cm long, narrowly lanceolate to linearly oblong, dense, especially at the stem base, rather long and conspicuous, deflexed. **Calyx**: segments shortly connate at the base, unequally 2-lobed, 0.7–1.0 cm, apically filiform. Bracts normally considerably exceeding the corolla, brown and papery, with conspicuous white hairs. **Corolla**: ± patent, tubular, curved, or nearly straight, glabrous or subglabrous surface, 1.0–1.5(–1.7) cm long; dark blood-red or purple, yellow at the base outside, shining dark red inside. Lower lip with small, subequal lobes; margins not ciliate. **Carpel**: stigma red to purple. **Stamens**: filaments glabrous above and usually also below, inserted 0.15–0.2 cm above the base of the corolla; anthers glabrous. **Host**: *Lotus cytisoides* (Fabaceae). **Ecology and distribution**: a coastal species, usually found on fixed dunes, rocky slopes and cliffs in the Mediterranean region: native to Algeria, Balearic Islands, Corsica, Croatia, east Aegean islands, Greece, Italy, Crete, Montenegro, Morocco, Portugal, Sardinia, Sicily, Spain and Tunisia. Flowering March–April. **Key diagnostics**: nearly uniform blood-red spikes of rather small flowers with red stigmas and conspicuous, long bracts are usually unmistakable. **Similar species**: misidentification with two morphologically and genetically similar species is possible: *O. foetida* and *O. densiflora*. *Orobanche variegata* has similar colouration but is often clumped, has foetid flowers with contrasting yellow stigmas and has different ecology (hill slopes, shrubby hosts).

Orobanche sanguinea
Crete, Greece (left, right, page 463)
Corsica, France (page 462)

NAME ***Orobanche santolinae*** Loscos & J.Pardo

SYNONYMS ***Orobanche artemisiae-campestris*** **subsp.** ***santolinae*** (Loscos & J.Pardo) O.Bolòs, Vigo, Masalles & Ninot; ***Orobanche loricata*** **subsp.** ***santolinae*** (Loscos & J.Pardo) O.Bolòs & Vigo; ***Orobanche major*** **subsp.** ***santolinae*** (Loscos & J.Pardo) R.C.V.Douin

Habit: slightly robust; inflorescence 8–17 × (2.1–)2.4–3.2(–3.9) cm, cylindrical, rounded or slightly conical and attenuated towards the apex, dense, sometimes somewhat lax at the base. **Height**: (13–)20–46 cm. **Stem**: somewhat robust, 4–8(–10) mm in diameter at its middle, bulbous at the base, villous-glandular (with long hairs up to 1.5 mm), frequently powdery in appearance when dry, purple or whitish slightly tinged with purple (light reddish-brown when dry). Basal scales 5–13 × (4–)6–10 mm, deltoid, ovate or oblong, imbricate; the uppermost (9–)12–20 × (3–)4–8 mm, broadly lanceolate, erecto-patent, villous-glandular, frequently powdery in appearance when dry. **Calyx**: 8–14 mm, segments free, contiguous, bipartite or sometimes subentire (both forms may occur on the same inflorescence), rarely bifid, teeth unequal sometimes subequal, subulate, rarely lanceolate, generally acuminate with a subfiliform apex, glandular-hairy, frequently powdery, whitish or yellowish in appearance with purple veins or sometimes completely purple. Floral bracts (8–)10–17(–19) × 4–7 mm, much shorter than (sometimes equalling) the lower lip of the corolla, rarely equalling the length of the corolla, broadly lanceolate to ovate, powdery when dry, purple or sometimes pink, soon turning brown. **Corolla**: 15–18(–20) mm, erect to erecto-patent, tubular, usually straight, rarely curved, slightly galeate at the apex, glandular-hairy, sometimes powdery when dry, ochroleucous or white with veins and apex tinged with purple. Upper lip emarginate to shortly bilobed, with patent lobes, rarely porrect; the lower lip with median lobe larger than the laterals, lobes sometimes subequal, obovate, deflexed, with margins irregularly crenate, glabrous or sometimes with glandular hairs. **Carpel**: stigma bilobed, purple, rarely orange or yellowish; style glandular-hairy. **Stamens**: filaments inserted obliquely, the adaxial ones inserted at (2–)3–4(–5) mm, the abaxial at 2–3(–4) mm from the base of the corolla; densely villous in the lower ⅔, sometimes only in the lower ½, glandular-hairy or with subsessile glandular hairs elsewhere; anthers hairy in the lower ½ of the suture. **Host**: exclusively on *Santolina* spp. (Asteraceae). **Ecology and distribution**: xerophytic scrub and rocky slopes at 20–1,900 m, in the western and central Mediterranean (Balearics, Corsica, France, Italy, Morocco and Spain). Flowering May–June (July). **Key diagnostics**: very polymorphic, chromatic variability considerable in a single population; subtle characteristics of the bracts, calyx and corolla, the indumentum, and host species. **Similar species**: easily-confused with other species in Subsection *Minores* of southern European coasts, many of them parasites of Asteraceae, including *O. artemisiae-campestris*, *O. cyrnea*, *O. apuana* and *O. calendulae*, but the relationships among these species require further research.

Orobanche santolinae
Spain (right) photo: José-Vicente Ferrández

NAME ***Orobanche schelkovnikovii*** Tzvelev

Habit: robust; inflorescence (15–)23–27(–43) cm, cylindrical, usually shorter than its stem, rarely equal to or longer, variably lax or dense. **Height**: (45–)50–65(–80) cm. **Stem**: glandular-pubescent, purplish, dirty pink or pale yellow-brown; scales (21–)24–30(–40) mm, broadly lanceolate to lanceolate. **Calyx**: segments broadly connate in front, with 2 lanceolate calyx-teeth, longer than ½ the corolla tube; (12–)15–16(–19) mm long. Floral bracts (16–)20–23(–26) mm, longer or rarely equal to the corolla, broadly lanceolate. **Corolla**: ± erecto-patent and medium-sized relative to plant; tubular to narrowly campanulate (considerably widened at the mouth); dorsal line slightly curved proximally, less curved or nearly straight in the middle portion and ± flexed forwards or straight distally (arcuate); (16–)22–25(–27) mm long, (8–)10–11 mm in diameter; pale pinkish-yellow with dull pale pink veins; externally glandular-pubescent with pale yellow hairs, interior glabrous or sparsely hairy on the upper lip. Upper lip sinuate, emarginate, with 2 broad lobes, these reflexed laterally; lower lip with 3 suborbicular lobes, irregularly dentate on the margins. **Carpel**: stigma lobes elongated-spherical to ovoid, yellow or bright yellow; style glandular-pubescent throughout. **Stamens**: filaments obliquely inserted at 2–3(–4) mm above the corolla base, sparsely hairy near the base, not glandular, the upper part below the anthers sparsely glandular-pubescent; anthers short-pubescent along sutures, especially in the ¾ part, rarely glabrous. **Host**: *Cirsium* spp. (Asteraceae), such as *C. aduncum*, *C. schelkovnikovii*, *C. cosmelii*, *C. leucocephalum* and *C. tomentosum*. **Ecology and distribution**: subalpine meadows, upper mountain belts, pastures, rocky and grassy slopes and valleys, at 2,000–2,300(2,600) m. Flowering July–August, fruiting August. Endemic to Armenia; possibly in adjacent regions. **Key diagnostics**: dirty pink, reddish, stout stems with few scales; dense, many-flowered inflorescences; calyx segments on side of bract broadly connate in front; host-specific. **Similar species**: *Orobanche kurdica* differs in its longer flowers, bracts, corolla curved at the base, higher filament insertion point, and free calyx segments, parasitic on Apiaceae; *O. grossheimii* differs in having dense, numerous, almost overlapping stem scales, shorter corolla that is curved at base, paler inside, with reddish to violet, rarely pale glands, calyx segments free, parasitic only on Caprifoliaceae.

Orobanche schelkovnikovii
Armenia (right, page 468, 469)

NAME ***Orobanche serbica*** Beck & Petrovič
SYNONYM ***Orobanche ozanonis*** F.W.Schmidt ex Beck

Habit: small, slender, few-flowered. **Height**: to 15 cm, rarely more. **Stem**: glandular-pubescent, yellowish or pinkish, slightly swollen at base; scales to 3 cm, lanceolate. **Calyx**: bifid with unequal segments, free or connate; apically filiform, brown and crisped; shorter than the corolla tube; 8 mm. Floral bracts lanceolate, deflexed, roughly equalling corolla. **Corolla**: erecto-patent, broadly cylindrical-campanulate with a rather open throat; dorsal line evenly curved; tube to 20 mm, pale yellow or whitish, sometimes tinged dull pink or violet distally and with darker veins; glandular-pubescent; upper lip subentire, forward-pointing; lower lip with equal, divergent lobes, rather sharply, irregularly toothed and with glandular-ciliate margins. **Carpel**: stigma lobes yellow, touching. **Stamens**: filaments usually hairy, inserted 3–4 mm above base of corolla; anthers sub-glabrous. **Host**: *Artemisia alba* (Asteraceae). **Ecology and distribution**: occurs in open scrub. Rather local, occurring in the Balkans, from Slovenia south to Albania, North Macedonia and western Bulgaria, to eastern Spain and southern France. Flowering (March) April–June. **Key diagnostics**: small, slender plant with campanulate corolla with purple-brown glandular hairs, ciliate margins and equal lower lip lobes on lower lip; host-specific. **Similar species**: similar to species in the complex Subsection *Minores*, differing in the characters stressed above.

Orobanche serbica
Albania (right)

NAME ***Orobanche staehelinae*** D.Pav., Michaud, Véla & J.-M.Tison

Habit: somewhat robust; entirely brownish-red; inflorescence short, never reaching ½ the total length of the plant (usually about ⅓ of the total length), rather dense, densely covered with short glandular hairs. **Height**: (12–)15–30(–40) cm tall. **Stem**: 7–8 mm in diameter; scales (16–) 17–20(–23) × (3–)4–5 mm. **Calyx**: segments (13–)15–17(–18) mm long, generally bidentate with unequal teeth often shorter than their tube, sometimes the longest equalling or exceeding the tube; segments rather broad and prominent. Bracts 17–20 × 5–6 mm, as long as, or shorter than the corolla. **Corolla**: 18–25(–26) mm long, erecto-patent after flowering, with a dorsal line straight in the middle (not uniformly curved); reddish with darker veins. **Carpel**: stigma dark honey- or chocolate-brown; style glabrous, rarely with few scattered hairs. **Stamens**: filaments inserted 3–4(–5) mm above the base of the corolla, hairy below, sparsely hairy along the dorsal line. **Host**: *Staehelina dubia* (Asteraceae). **Ecology and distribution**: known only from south-eastern France, on open, calcareous, rocky scrub, at an elevation of 140–500 m. Blooming in late May and early June. **Key diagnostics**: distinguished by its brownish-red colouration, short, dense inflorescences, rather prominent calyx, and its host. **Similar species**: morphologically close to *O. loscosii* and *O. centaurina*, differing mainly in the colour of the corolla and stigma, and the host.

Orobanche staehelinae France (above, right, pages 474, 475) photos: Daniel Pavon

NAME ***Orobanche subbaetica*** Triano & A.Pujadas

Habit: spike dense with many flowers, laxer below. **Height**: up to 42(–56) cm. **Stem**: robust, purple at least above, and with long glandular hairs throughout. Scales to 15(–25) mm long, triangular-lanceolate. **Calyx**: to 16 mm long, with free, ± unequal 2-toothed or entire segments, purple, paler towards the base and deeper towards the teeth, glandular-hairy throughout. Floral bracts to 18 mm, equalling or just exceeded by the corolla. **Corolla**: erecto-patent with a smooth dorsal line, 18–22 mm long, tubular. Upper lip deeply bilobate, with lobes erect or bent forward during the anthesis; dorsally purplish-blue and ventrally whitish; lower lip with deflexed, distant, rectangle-shaped and marginally ± crenate-denticulate lobes. **Carpel**: stigma dark purple, glabrous, lobes just touching. **Stamens**: filaments inserted 2–3 mm from the base, hairless or virtually so. **Host**: known only to grow on *Antirrhinum* (*A. australe* and *A. litigiosum*), Plantaginaceae. **Ecology and distribution**: rather rare and strongly southern in distribution: Gibraltar, Morocco (to be confirmed) and Spain. Flowering in May and early June. **Key diagnostics**: corolla dorsally purplish-blue and ventrally whitish with short, rectangular and divergent lower lip lobes; stigma deep blackish-purple; host-specific. **Similar species**: belonging to the similar and complex Subsection *Minores*: differs from *O. minor* in its denser, more robust spikes, stigma colour and other characters stressed above, from *O. calendulae* in having erecto-patent, purplish flowers with uniformly curved corolla dorsal line, and from *O. amethystea* in the lack of geniculate corolla dorsal line and the lack of filiform calyx-teeth. Specificity for *Antirrhinum* is the clearest diagnostic.

Orobanche subbaetica Spain (above, right) photos: Antonio Pujadas Salvá

NAME ***Orobanche teucrii*** Holandre
SYNONYM ***Orobanche atropurpurea*** F.W.Schultz

Habit: slender; inflorescence usually short and lax with few flowers. **Height**: (15–)20–25(–40) cm. **Stem**: slender and erect, light brown, reddish or yellow-brown, glandular-pubescent; scales 10–15 mm, triangular to ovate and lanceolate. **Calyx**: segments usually bifid to the middle, with almost equal halves, about ⅓ or ½ as long as the corolla, glandular-pubescent; 10–14 mm. Floral bracts 15–23 mm, lanceolate, equalling or shorter than the corolla, glandular-pubescent. **Corolla**: erect, narrowly infundibuliform, dorsal line slightly curved in the lower part, almost straight in the middle and strongly humped behind the upper lip; 18–22(–25) mm; brownish-yellow, brownish-violet, or reddish with darker veins. Upper lip almost entire or slightly emarginate, with raised lobes and glandular hairs; lower lip lobes almost equal, usually glabrous, crenate, deflexed. **Carpel**: stigma lobes divergent, dark brown, purple, reddish, orange, rarely yellow; style markedly glandular-pubescent. **Stamens**: filaments inserted 3–5 mm above corolla tube. Filaments are densely glandular-pubescent. Anthers usually pubescent at the line of fusion. **Host**: exclusively on *Teucrium* (Lamiaceae), especially *T. chamaedrys*, *T. montanum* and *T. pyrenaicum*. **Ecology and distribution**: xerothermic grassland, rocky slopes and scree. Flowering usually May–June (July). Mainly a central European species, occurring from the Pyrenees to the Carpathian Mountains, north to Belgium, south to north-eastern Spain, Italy, the northern Balkans and Romania. **Key diagnostics**: usually a small plant, with short, lax, few-flowered inflorescences; corolla erect, narrowly infundibuliform, brownish-yellow or brownish-violet with dark stigma; host-specific. **Similar species**: similar to the more widespread *O. caryophyllacea*, but the latter is larger with erecto-patent and fragrant flowers, broadly campanulate corolla, considerably widened at the mouth, with a dorsal line evenly curved from the base, broad upper lip, lower filament insertion point, and different host range (Rubiaceae).

Orobanche teucrii
Czech Republic (left, right)
photos: František Lamla
Austria (page 480, 481)

NAME ***Orobanche variegata*** Wallr.

SYNONYMS ***Catodiacrum variegatum*** (Wallr.) Dulac; ***Orobanche condensata* subsp. *variegata*** (Wallr.) Batt.; ***Orobanche foetida* subsp. *variegata*** (Wallr.) Ball; ***Orobanche spartii*** Guss.

Habit: clumped, often with many stems ascending from the base. **Height**: 30–70 cm. **Stem**: rather slender, 0.4–1.0 cm wide, swollen at the base, glandular-pubescent, yellow or reddish-maroon; scales 1.5–3.0 cm, oblong ovate to linear-lanceolate. **Calyx**: segments unequal, triangular-lanceolate, subulate; much shorter than the corolla tube, 1.0–2.0 cm, apically crisp and blackish. **Corolla**: a short, wide tube, rather smoothly arched along the dorsal line; 14–23 mm long, brownish-red to maroon (rarely yellow); ± concolourous inside and out; upper lip markedly 2-lobed; lower lip with a middle lobe, twice as big as the laterals; margins entire. **Carpel**: stigma lobes yellow, divergent; style glabrous below and glandular-pubescent below stigma. **Stamens**: inserted 2–4 mm, filaments villous below; anthers pubescent at the line of fusion. **Host**: woody Fabaceae: *Calicotome* spp., *Spartium junceum*, *Genista* spp. and *Cytisus* spp. **Ecology and distribution**: central Mediterranean: Italy, Sicily, Sardinia, Corsica, southern France, Tunisia and Algeria; common in Sicily and in the Aeolian islands. Occurs on medium to high elevation hillsides and scrub. Flowering (mid-April) May–June (July). **Key diagnostics**: dark red colouration; clumped plants with foetid flowers, corolla concolourous inside and out with contrasting yellow stigmas, the middle lobe of the lower lip is about twice the size of the side lobes. **Similar species**: several species possess dull red flowers with yellow stigmas: *O. variegata* is similar in appearance to *O. sanguinea*, which overlaps in distribution but has scentless flowers with red (not yellow) stigmas, and differs in host range and ecology; *O. variegata* has foetid flowers with a wider, campanulate corolla, protracted lower lip and villous filament bases. Relationships among these two taxa, together with the similar more westerly species, *O. foetida*, need further investigation. *Orobanche variegata* can also be confused with *O. gracilis*, which differs particularly in having subequal lower lip lobes.

Orobanche variegata Sicily, Italy (above, right, page 484) photos: Uwe Grabner, Sardinia, Italy (page 485) photo: Uwe Grabner

NAME ***Orobanche zajaciorum*** Piwow.

Habit: slender; inflorescence (2.8–)4.0–8.5 cm, cylindrical to slightly ovoid, shorter than its stem, ± dense. **Height**: (10–)15–17(–24) cm. **Stem**: slender, densely glandular-pubescent in the upper part, pale dull pink violet or yellowish; scales 10–13(–15) mm, ovate-lanceolate (triangular) to lanceolate. **Calyx**: segments simple, free, entire, with long acuminate teeth and rather ovate base; (5–)7–8(–11) mm long, shorter than the corolla tube. Floral bracts 8–13 mm, shorter or equal to the corolla, narrowly lanceolate, glandular-pubescent. **Corolla**: strongly cernuous, appearing arched downward; tubular to narrowly campanulate, the dorsal line strongly geniculate, scarcely inflated at base; flowers rather small, (11–)13–15(–17) mm long, 3–4 mm in diameter in the central part; tube basally yellowish-white, distally pale pink with violet or dark pink veins, externally glandular-pubescent. Upper lip with 2 short, broad lobes, emarginate, patent; lower lip with 3 ovate lobes, the central lobe slightly larger than the laterals, the margins irregularly dentate. **Carpel**: stigma lobes dark pink, violet to purple; hidden in the corolla, not protruding; style with very sparse glandular hairs or glabrous. **Stamens**: filaments obliquely inserted 4–5 mm above the corolla base, densely villous around the base and middle with non-glandular and glandular hairs; distally with sparse, short, glandular hairs; anthers glabrous, with sparse short glandular hairs at the line of fusion. **Host**: *Scutellaria* spp. (Lamiaceae), especially *S. orientalis*, *S. sosnowskyi* and *S. karjaginii*. **Ecology and distribution**: rocky slopes and steppe, at 1,000–1,700 (2,000) m. Flowering end May–June, fruiting (June) July (August). Probably endemic to the Caucasus, in Georgia, Armenia and Russia (Republic of Adygea). **Key diagnostics**: unmistakable for its pinkish, distinctively curved, geniculate, small flowers, simple and narrow calyx segments, strong carnation scent, dark pink, concealed stigma; host-specific. **Similar species**: somewhat similar to *O. caryophyllacea* and *O. lutea*; both are much taller with larger flowers (18–30 mm), broad bidentate calyx segments, and different hosts; *O. lutea* has a yellow stigma.

Orobanche zajaciorum
Georgia (left, right, page 488)
Armenia (page 489) photo: Zbigniew Iwański

NAME ***Phacellanthus tubiflorus*** Siebold & Zucc.

Habit: low, solitary, often clustered inflorescences, each with 4–6 flowers, at ground-level. **Height**: (4–)5–11 cm. **Stem**: unbranched, glabrescent; scales ovate-triangular, 5–8(–10) × 3–4 mm with slightly membranous margins and acute apex; scales 1.5–2.3 cm × 5–8 mm, with acuminate or sub-obtuse apex. **Calyx**: 2–3 free, narrowly linear, caducous calyx lobes, resembling bracteoles. Flower bracts ovate to lanceolate, obtuse or acute at apex, usually shorter than corolla tube. **Corolla**: tubular, 2.5–3.5 cm, white or pale yellow with subequal oblong lobes, each ca. 3 × 2 mm; upper lip orbicular, to 4 mm; lower lip deeply trilobed, to 3 mm. **Carpel**: stigma pale; style glabrous, 13–16(–20) mm. **Stamens**: filaments 12–14 mm, sparsely pubescent near insertion point; anthers ca. 1.8 mm long, ovate, pale orange. **Hosts**: *Actinidia* (Actinidiaceae), *Fraxinus* (Oleaceae), *Machilus* (Lauraceae), *Ulmus* (Ulmaceae), *Cornus* (Cornaceae), *Kalopanax* (Araliaceae) and *Picrasma* (Simaroubaceae). **Ecology and distribution**: mixed to broadleaf humid temperate forests across eastern China, Japan and Korea, and north to Pacific south-east Russia. Flowering July–August. **Key diagnostics**: whitish clusters of flowers at ground-level on the forest floor; lack of functional calyx. **Similar species**: unlikely to be confused with any other species; *Christisonia hookeri* is superficially somewhat similar but has a defined calyx and yellow markings on the corolla interior.

Phacellanthus tubiflorus
Japan (right)
Japan (page 492, 493) photos: Kenji Suetsugu

NAME ***Phelipanche aedoi*** Carlón, G.Gómez, M.Laínz, Moreno Mor., Ó.Sánchez & Schneew.

Habit: tall, robust and dense spikes with many blue flowers, often appearing from near ground-level, on pale yellow stems. **Height**: to 35(–45) cm. **Stem**: yellowish-cream, glandular-hairy; scales to 10 mm long, triangular to triangular-lanceolate. **Calyx**: to 12 mm long with 4 narrowly triangular teeth, equalling or exceeding the tube; floral bracts to 12 mm, triangular-lanceolate, equalling or shorter than the calyx; bracteoles linear-lanceolate and much shorter than the calyx; bracts, bracteoles and calyx lobes cream, concolourous. **Corolla**: ± erect, tubular-funnel-shaped with a short tube to 19(–22) mm long, glandular-hairy, pale blue except for the very base (where whitish); throat with whitish folds; upper lip bilobed with erect, entire, divergent lobes with a clear sinus; lower lip with deflexed subequal lobes, somewhat overlapping. **Carpel**: stigma white, hairless, 2-lobed. **Stamens**: inserted 2.8–5 mm above the corolla tube base, slightly hairy at the base with non-glandular hairs; anthers virtually glabrous. **Host**: mainly on *Launaea arborescens* and, to a lesser extent, *Sonchus tenerrimus* (Asteraceae). **Ecology and distribution**: rare and local; parasitic on desertic shrubs in the Almería desert (south-eastern Spain). **Key diagnostics**: long, dense, many-flowered spikes, parasitic primarily on *Launaea* in semi-deserts in southern Spain. **Similar species**: somewhat similar to *P. mutelii* and *P. nana* but far more robust than both, and with more numerous flowers; distinguished from *P. schultzii* by its shorter, not subulate calyx-teeth and corolla lobes with entire margins.

Phelipanche aedoi
Spain (right) photo: Gonzalo Moreno Moral

NAME ***Phelipanche aegyptiaca*** (Pers.) Pomel
SYNONYMS ***Kopsia aegyptiaca*** (Pers.) Caruel; ***Orobanche aegyptiaca* var. *lasianthera*** Maire; ***Orobanche longiflora*** Pers.; ***Phelypaea aegyptiaca*** (Pers.) Dumort.

Habit: clumped with numerous ascending stems. **Height**: 15–30 cm. **Stem**: usually ± robust, branched (branches erect, clustered, almost parallel to the stem); sometimes simple. **Calyx**: teeth narrowly triangular-subulate, usually longer than their tube (sometimes, equalling the tube); 9–10 mm. Floral bracts 7–10 mm. Lower flowers pedunculate. **Corolla**: sub-erect, infundibuliform; upper lip deeply 2-lobed, lobes ± erect, rounded; lower lip deeply 3-lobed, lobes ovate to obovate, obtuse; 20–22(–37) mm; pale violet with darker lobes. **Carpel**: stigma white. **Stamens**: anthers hairy. **Hosts**: numerous herbs, mainly cultivated crops in the Apiaceae, Brassicaceae, Cucurbitaceae, Fabaceae, Malvaceae and Solanaceae. **Ecology and distribution**: cultivated fields; widespread across the Old World; common in much in northern Africa, Asia and the Middle East, primarily in the subtropical biome. **Key diagnostics**: branched spikes with rather large, infundibuliform flowers and hairy anthers, on cultivated hosts. **Similar species**: *Phelipanche ramosa* and *P. nana* are similar but smaller in all parts, with corolla 10–15(–17) mm and sub-glabrous anthers. The poorly known *P. brassicae* (Novopokr.) Soják (previously described as *P. mutelii* subsp. *brassicae*), reported mainly from south European Russia and Ukraine, is a parasite of cultivated crops such as cabbage, tomato, rape and mustard, and is characterised by its pale purple corolla (20–22 mm long) and glabrous anthers; its taxonomic position and relationship with *P. aegyptiaca* and *P. ramosa* require further investigation.

Phelipanche aegyptiaca Georgia (above, right)

NAME ***Phelipanche arenaria*** (Borkh.) Pomel
SYNONYMS ***Orobanche arenaria*** Borkh.; ***Orobanche artemisae-vulgaris*** F.W.Schultz; ***Orobanche laevis*** L. nom. rej.; ***Kopsia arenaria*** (Borkh.) Dumort.; ***Phelypaea arenaria*** (Borkh.) Walp.; ***Phelypaea comosa*** G.Don

Habit: usually robust; inflorescence cylindrical, usually many-flowered; stems often clumped. **Height**: 15–60 cm. **Stem**: robust or rarely slender, erect, rarely branched, yellow, yellowish-brown or pale violet, glandular-pubescent, with numerous, triangular-lanceolate to lanceolate scales up to 20(–25) mm long. **Calyx**: tubular-campanulate; 4–5-dentate, teeth narrow-lanceolate and subulate, longer than or sometimes equalling the tube, glandular-pubescent; 11–18(–21) mm; bracteoles lanceolate, ± about the length of the calyx segments, glandular-pubescent. Bracts lanceolate, ± ½ the width of the corolla, glandular-pubescent. **Corolla**: erect to erecto-patent, ± tubular-infundibuliform, campanulate, usually without darker longitudinal nerves; corolla dorsal line ± evenly curved from the base onwards, almost straight in the middle and sometimes bent upwards at the tip; (18–)21–35(–38) mm; blue, pale blue, violet, blue-violet. Upper lip 2-lobed, lobes erect and pubescent; lower lip with 3 lobes of almost equal size, orbicular or broadly ovate, rarely acute, plicate, deflexed. **Carpel**: stigma white; style with many glandular hairs. **Stamens**: filaments inserted 5–7 mm above the base of the corolla tube; sub-glabrous at the base, or sparsely glandular below; anthers ± densely hairy to woolly. **Host**: *Artemisia* spp. (Asteraceae), mainly *A. campestris*; less often *A. lercheana* or other species. **Ecology and distribution**: xerothermic grasslands, steppe, rocky, loamy and sandy slopes, ecotone fields, ruderal sites. Flowering June–July, fruiting July–August. Widespread from north-west Africa (Algeria and Morocco) and Europe to the Caucasus, Asia Minor and central Asia (Kazakhstan, north-east of the Caspian Sea). **Key diagnostics**: yellowish stems with numerous, triangular-lanceolate to lanceolate bracts; erect to erecto-patent flowers, corolla widely funnel-shaped in the upper part; anthers woolly; mainly parasitic on *Artemisia campestris.* **Similar species**: *Phelipanche purpurea* has a slenderer stem with sparse lanceolate bracts; flowers erecto-patent to patent, smaller, usually with darker longitudinal nerves; anthers not woolly; calyx shorter, equalling or exceeded by the tube; stem violet to blue-grey. *Phelipanche caesia* has patent (usually forward-curved) and densely white arachnoid-villous to lanate corolla, and glabrous anthers.

Phelipanche arenaria
Czech Republic (left, right)
photos: František Lamla
Poland (page 500, 501)

NAME ***Phelipanche bohemica*** (Čelak.) Holub & Zázvorka
SYNONYMS ***Orobanche bohemica*** Čelak.; ***Phelipanche purpurea* subsp. *bohemica*** (Čelak.) Zázvorka

Habit: slender to stout; inflorescence dense, usually many-flowered, up to 27(–35) cm long and 20–60 flowers, elongated, cylindrical. **Height**: (15–)25–50(–70) cm. **Stem**: usually stout, unbranched, violet, blue-grey or yellowish, glandular-pubescent, with sparse, lanceolate to narrowly lanceolate scales. **Calyx**: campanulate, dark violet, with 5 calyx-teeth (3–)4–7 mm long, 8–12 mm long in total. Floral bracts lanceolate. **Corolla**: erecto-patent, ± tubular to narrowly campanulate, somewhat widened towards the throat, with a curved dorsal line, not bent; 20–30 mm, intense blue-violet. Upper lip with 2 erect violet-veined lobes; lower lip with 3 ovate, obtuse lobes with margins somewhat irregularly shallowly lobed. **Carpel**: stigma lobes contiguous, yellowish; style glandular-pubescent. **Stamens**: filaments inserted 3–5 mm above base of the corolla tube; sub-glabrous or sparsely glandular at the base; anthers glabrous. **Host**: exclusively on *Artemisia campestris* (Asteraceae). **Ecology and distribution**: xerothermic grassland, fallow lands, field edges and sunny slopes. Flowering usually June–July. Confirmed localities known from central and south-western Europe, Germany, Poland, the Czech Republic, France, Austria, Switzerland, Italy and Spain. Formerly included under *P. purpurea* s.l., so further work necessary. **Key diagnostics**: tall, with many flowers, almost entirely dark blue-violet in colour with contrasting yellow stigmas; host-specific. **Similar species**: differs from *P. purpurea* s.str. in its taller and stouter stem, longer and denser inflorescence; longer, erecto-patent flowers, somewhat widened towards the throat, with a curved dorsal line; yellowish stigma; more intensely coloured in all parts: with dark violet; ovate, obtuse lobes with distinctive 'eroded' margins; anthers glabrous. Always on *Artemisia campestris*.

Phelipanche bohemica
Poland (right)

NAME ***Phelipanche caesia*** (Rchb.) Soják
SYNONYMS ***Orobanche caesia*** Rchb.; ***Orobanche borealis*** Turcz.; ***Phelypaea caesia*** (Rchb.) Griseb.; ***Phelypaea lanuginosa*** C.A.Mey.

Habit: somewhat robust; inflorescence (6–)8–15(–22) cm, ovoid to cylindrical with rounded apex, lax to dense, much shorter than its stem. **Height**: (12–)16–25(–35) cm. **Stem**: simple, thick, usually robust, yellowish-brown, violet to bluish, pubescent with short hairs or white-woolly; scales broadly lanceolate and glandular-pubescent, ± densely white arachnoid-villous to lanate. **Calyx**: tubular-campanulate, segments 4–5, narrowly lanceolate, acuminate, almost equalling tube or shorter, glandular-arachnoid-pilose; 8–12(–18) mm. Floral bracts 8–12(–18) mm, lanceolate to linear-lanceolate, almost equalling calyx, abundantly glandular-arachnoid-pilose. **Corolla**: tubular, erect or erecto-patent at base and ± horizontal or distinctly forward-inflected distally, slightly contracted below and funnel-shaped in the upper ½; dorsal line ± evenly curved; (15–)17–19(–25) mm; sky-blue to pale violet, occasionally with ± darker nerves; usually densely hairy externally, glabrous inside except the folds, and with sparsely hairy lobes. Upper lip deeply bilobate, lobes oblong to ovate, obtuse, rarely subacute, with crenate and ciliate edges; lower lip lobes ovate, elongated, crenate, almost equal. **Carpel**: stigma lobes spherical, white or rarely yellowish-white; style sparsely glandular-pubescent (hairs more abundant beneath the stigma). **Stamens**: filaments inserted 6–8 mm above the corolla base; glabrous or rarely slightly hairy at the base; anthers glabrous or rarely along sutures with sparse long hairs. **Host**: *Artemisia* spp. (Asteraceae), especially *A. austriaca*, *A. fragrans*, *A. pontica* and *A. absinthium*. **Ecology and distribution**: steppe, xerothermic grassland, dry hills, sandy and rocky slopes, roadsides, ruderal habitats, open forests and scrub. Flowering late May–July (August). Central Europe (rare except the south-east) to Siberia, the Caucasus and central Asia. **Key diagnostics**: unbranched, woolly stems; inflorescence with rounded apex; flowers patent (usually curved forward), densely white arachnoid-villous to lanate; floral bracts equalling the calyx. **Similar species**: other pubescent, unbranched species include *P. pulchella*, *P. hajastanica* or *P. zangezuri*. *Phelipanche caesia* differs from these in the characters stressed above, and the host. *Phelipanche pallens* (Bunge ex Ledeb.) Soják, a poorly known species, also reported to grow on *Artemisia*, differs in its absence of woolly pubescence in the inflorescence, usually hairy anthers, and more easterly range. Potentially confused with *Orobanche coerulescens*, another whitish-woolly species and parasite of *Artemisia*, but this species lacks the characteristic bracteoles of *Phelipanche*.

Phelipanche caesia
Czech Republic (left, right)
photos: František Lamla
Armenia (page 506, 507)

NAME ***Phelipanche camphorosmae*** Carlón, G.Gómez, M.Laínz, Moreno Mor., Ó.Sánchez & Schneew.

SYNONYM ***Orobanche camphorosmae*** (Carlón, G.Gómez, M.Laínz, Moreno Mor., Ó.Sánchez & Schneew.) A.Pujadas & Triano

Habit: typically stunted spikes with flowers from ground-level. **Height**: 13–25.5 cm. **Stem**: simple (or branched from the very base); creamy yellow, whitish or violet-tinged, with short glandular hairs 0.4 mm long. **Calyx**: segments triangular, creamy yellowish, violet when dry, densely covered in glandular hairs; 8.3–11 mm. Floral bracts (5.5–)7.4–11.2 × (2.1–)2.5–3.8 mm. **Corolla**: tubular rather than funnel-shaped; 17.8–21.5 mm in length, 2.5–3 mm in diameter at the base and 4.4–5.1 mm at throat; pale blue or violet apart from the basal area below the stamen and the ventral folds where it is whitish (folds woolly); veins inconspicuous; upper lip with 2 erect ovate lobes with long non-glandular hairs; lower lip with 3 obovate-spathulate or ovate lobes shaped with hairs similar to those of the upper lip. **Carpel**: stigma white or cream; style white, sometimes violet-tinged, with short hairs beneath the stigma. **Stamens**: filaments inserted 4–6 mm above the corolla tube base, generally white, non-glandular-hairy at the base. **Host**: *Camphorosma monspeliaca* (Amaranthaceae). **Ecology and distribution**: found on coastal turf and dunes in Spain and southern France. Flowering May–June. **Key diagnostics**: short, stunted spikes with few flowers; host specificity for *Camphorosma*. **Similar species**: differs from the more widespread *P. mutelii* in its rather tubular, scarcely infundibuliform flowers (not widened in the throat), usually paler with nerves hardly marked.

Phelipanche camphorosmae
Spain (right)
photo: Gonzalo Moreno Moral

NAME ***Phelipanche cernua*** Pomel
SYNONYMS ***Orobanche inexspectata*** (Carlón, G.Gómez, M.Laínz, Moreno Mor., Ó.Sánchez & Schneew.) Domina, Greuter, P.Marino & P.Schäf.; ***Phelipanche inexspectata*** Carlón, G.Gómez, M.Laínz, Moreno Mor., Ó.Sánchez & Schneew.

Habit: slender; inflorescence lax, relatively few-flowered, mostly oblong-ovoid, often elongated (equal or sometimes much longer than its stem). **Height**: 10–30 cm. **Stem**: relatively slender with few scales; simple or rarely with few erect, short branches, ash-blue with a greyish pale blue, rarely yellowish-brown hue, glandular-pubescent. **Calyx**: segments narrowly triangular but not distally filiform, grey-violet, rarely cream, subtly veined; teeth 4–9 mm long (equal to the tube or sometimes longer), much shorter than the corolla, 9–15 mm. Floral bracts ½ the length of the corolla or a little longer, grey-blue, sometimes yellowish-brown. **Corolla**: erecto-patent to patent, ± narrowly tubular to slightly infundibuliform, constricted and ± geniculate above the ovary, straight along the dorsal line, curved, then ± straight again, sometimes slightly curved at the tip; veined and glandular-pubescent; 15–20 mm; pale grey-blue, ± deep blue-violet, rarely blue-cream; upper lip lobes usually acute, deflexed and pubescent; lower lip lobes rounded to oblongoid, convergent and entire, pubescent, with white folds. **Carpel**: stigma lobes yellow. **Host**: exclusively *Lactuca* spp., mainly *L. viminea* and *L. quercina* subsp. *wilhelmsiana* (Asteraceae). **Ecology and distribution**: xerothermic habitats, rocky and loamy slopes, steppe, ruderal habitats and roadsides. Flowering end May–June (July), fruiting July–August. Mediterranean Basin (Italy, Algeria, France, Morocco and Spain) and western Asia (Armenia and Azerbaijan); probably more widespread in the form of *P. cernua* subsp. *sinaica* (*P. sinaica*) in Egypt, Syria, Afghanistan, Iran and Azerbaijan; the distinctiveness of this taxon and its distribution requires further research. **Key diagnostics**: distinctive ash-blue, lax inflorescence, somewhat humped and ± geniculate corolla, yellow stigma, calyx-teeth much shorter than corolla; host-specific. **Similar species**: *Phelipanche schultzii* has a taller and stouter, longer inflorescence, calyx-teeth much longer than their tube (almost equalling the corolla), purple or violet corolla, divergent and regularly toothed or apiculate lower lip lobes, and white (or cream) stigma. It can be somewhat similar to *P. rosmarina*, but its morphological and ecological characteristics are distinct.

Infraspecific taxa

Subsp. *cernua* is the widespread form with rather long flowers with larger corolla lobes, and triangular-lanceolate to lanceolate calyx-teeth, longer than the calyx tube. Widespread across the Mediterranean Basin and western Asia.

Subsp. *sinaica* has a shorter corolla with smaller lobes, calyx-teeth shorter than their tube, shortly triangular and acute. Strongly southern and eastern: Egypt, Syria, Afghanistan, Iran and Azerbaijan.

Phelipanche cernua
Armenia (right)

NAME ***Phelipanche chionistrae*** Rätzel, Hand, Christodoulou & Uhlich

Habit: short, slender spikes with few flowers ± from ground-level. **Height**: 6.5–8 cm. **Stem**: simple, with few scales, densely glandular-hairy with yellow glands; rather dark blackish or brownish-violet to dull, dark red, 1.2–3 mm in diameter. **Calyx**: with 4 teeth approximately 6–7 mm long, concolourous with stem; narrowly triangular-acuminate and densely glandular-pubescent, 6–7 mm. Floral bracts 5–7 × 3–4 mm, concolourous with stem, broadly triangular-acuminate, whitish glandular-hairy. **Corolla**: narrowly tubular with a slight curve above the constriction; 10–13 mm; dark purplish-violet with a whitish base concealed within the calyx tube; tube with darker veins and numerous light glandular hairs. Upper lip bilobed, rectangular and vertically erect; lower lip with 3 lobes of equal size, rounded. **Carpel**: stigma lobes pale yellow; style densely hairy beneath the stigma. **Stamens**: filaments inserted 3–3.5 mm above the base of the corolla tube, whitish hairy beneath, glabrous above; anthers white, with sparse hairs. **Host**: *Odontarrhena troodi* and possibly *O. cyprica* (Brassicaceae); possibly also *Teucrium cyprium* (Lamiaceae) and *Hypericum stenobotrys* (Hypericaceae). **Ecology and distribution**: native to Cyprus: endemic to high altitude, bare slopes of Mount Chionistra. Flowering April–June. **Key diagnostics**: dark, almost blackish, stunted spikes with few flowers; unique ecology. **Similar species**: most similar to *P. rosmarina* and *P. olbiensis*, differing in colouration, calyx morphology, indumentum, host, distribution and ecology.

Phelipanche chionistrae
Cyprus (left, right)
photos: Charalambos S. Christodoulou
(*Flora of Cyprus*)

NAME ***Phelipanche cilicica*** (Beck) Soják
SYNONYM ***Orobanche cilicica*** Beck

Habit: slender; inflorescence ovoid or short-cylindrical, usually few-flowered, shorter than or equalling its stem, up to 8 cm long. **Height**: 8–12(–20) cm. **Stem**: yellowish-brown, usually unbranched (rarely short-branched), with few, 8–10 mm long, usually lanceolate scales, glandular-pubescent. **Calyx**: teeth shorter than or rarely equalling their tube, glandular, deltoid to lanceolate at the base, subulate at the tip; 8–14 mm. Floral bracts broadly lanceolate, acuminate, 6–10 mm long, equalling the calyx or shorter, glandular-pubescent. Bracteoles linear-subulate, shorter than the calyx. **Corolla**: erecto-patent, rarely slightly inclined, abundantly glandular-pilose; tubular-infundibuliform, dilated close to the base and distinctly bent forward at the limb; 18–24 mm; pale-bluish, yellow to yellowish-white, glandular-pubescent outside. Upper lip lobes with fimbriate, entire or weakly crenate margins; lower lip lobes broadly ovate, folds glandular-pubescent. **Carpel**: stigma usually white; style sparsely glandular-pubescent. **Stamens**: filaments with few glandular hairs; anthers sparsely hairy at the base. **Host**: mainly on *Stachys inflata*, rarely *Phlomis orientalis* (Lamiaceae); occasionally on *Turgenia latifolia* (Apiaceae); reports on *Salvia* spp. need verification. **Ecology and distribution**: semi-arid sandy and rocky slopes, dry hills, semi-deserts, steppe. Asia Minor, Armenia–Kurdistan, Armenia, Azerbaijan, Turkey and Iran. **Key diagnostics**: few-flowered spikes with pale bluish or yellow to yellowish-white, abundantly glandular-pilose flowers; corolla tubular-infundibuliform; calyx-teeth shorter than their tube; parasitic on Lamiaceae. **Similar species**: *Phelipanche pulchella* differs in its infundibuliform corolla gradually broadened towards the limb, usually darker blue or violet and lanate, with characteristic longer, stiff hairs, and calyx-teeth equalling or slightly longer than their tube. *Phelipanche nana* has smaller and sparser hairs on the corolla, which is smaller, 11–15(–17) mm, and calyx-teeth equalling or longer than their tube. *Phelipanche cilicica*, *P. pulchella* and *P. nana* as a group are all somewhat similar to the poorly known taxon *P. kelleri* (Novopokr.) Soják, described from Kazakhstan, a parasite of *Bassia prostrata* (Amaranthaceae), and also reported from the south and east of European Russia and the Caucasus; this plant is characterised by its dense inflorescence and by its corolla dimensions (18–20 mm long), and by completely glabrous lower lip folds.

Phelipanche cilicica Armenia (above, right)

NAME ***Phelipanche cingularum*** Croze, Carlón, J.M.Tison, Michaud, J.Molina & Moreno Mor.

Habit: rather stout, simple or sparingly branched; inflorescence long, occupying at least ½ its stem, rather lax, with flowers erecto-patent. **Height**: up to 25 cm. **Stem**: cream-coloured, sometimes slightly tinged orange-brown at the apex, simple or with few short branches arising from the lower ⅓, densely covered with white, retrorse hairs; scales few, lanceolate to narrowly triangular. **Calyx**: weakly pigmented, cream to pale orange-brown, with short subulate teeth. Bracteoles less than ½ the width of the corolla. Floral bracts similar to the scales, lanceolate or narrowly triangular, much shorter than their flower, initially cream-coloured, later with a crisp, brownish apex. **Corolla**: 20–27 mm long, tubular-infundibuliform, densely hairy both internally and externally, with rather long glandular trichomes, light blue with darker longitudinal veins, dorsally markedly bent at ca. ⅓ of its length, then uniformly curved with no subapical hump, ventrally straight. Upper lip patent, bifid, with triangular, acute, irregularly dentate, conspicuously ciliate lobes; lower lip reaching ⅓ of the corolla length, trifid, with poorly divergent, spathulate, dentate, often apiculate lobes, and 2 prominent, whitish, densely hairy bumps between the lobes and the tube. **Carpel**: stigma slightly bilobed, white; style white, increasingly bluish towards its apex, sparsely hairy throughout. **Stamens**: filaments inserted at the level of the lower angulation of the corolla, ca. ⅓ of its length; densely covered with long hairs, distally bluish-white and sparsely hairy; anthers mostly glabrous. **Host**: Brassicaceae, mainly *Hesperis laciniata*, rarely *Biscutella lima*, *Pseudoturritis turrita* and *Erysimum* spp. **Ecology and distribution**: a rare and local species known from the Mediterranean limestone mountains of southern France, at the foot of cliffs or on ledges, at elevation 170–1,400 m. Flowering in May and early June. **Key diagnostics**: peculiar ecology; stems rather stout, simple or sparingly branched, usually light-coloured; inflorescence elongated, lax, with erecto-patent flowers; corolla relatively large with acute lobes. **Similar species**: *Phelipanche gussoneana* has a far smaller and distinctly infundibuliform corolla, with entire lobe edges, almost glabrous filaments, and denser and shorter inflorescences, and is more strongly pigmented throughout.

Phelipanche cingularum
France (right, page 518, 519)
photos: Thomas Croze

NAME ***Phelipanche daninii*** (Domina & Raimondo) Domina

Habit: usually robust, clumped with variably sized stems. **Height**: 15–50 cm. **Stem**: simple, or branched (from the base), slender, glandular-pubescent; pale brown. **Calyx**: densely glandular-pubescent with segments lanceolate-subulate, light yellow, shorter than to equalling the tube; calyx-teeth 5–10 mm long. Floral bracts 6–8 mm, ovate-lanceolate, shorter than calyx. Bracteoles linear-subulate. Lower flowers sessile. **Corolla**: straight or slightly curved, infundibuliform in the distal part constricted above the ovary, inflated near the upper lip; 16–22 mm; pale bluish to violet distally, with darker veins, cream proximally. Upper lip bilobed, folded, ± ovate with margins entire; lower lip lobes ovate, elongated, entire, almost equal size, deflexed. **Carpel**: stigma lobes whitish or bluish. **Stamens**: filaments inserted 4–7 mm above corolla base, densely hairy below, hairy above; anthers glabrous. **Host**: parasitic on *Moricandia nitens*, possibly other Brassicaceae. **Ecology and distribution**: restricted to mudstone hills and desert flats in the Middle East. Flowering period February–March. Endemic to southern Israel and southern Jordan. **Similar species**: differs from *P. aegyptiaca* mainly in that its flowers are not pedunculate, and in its smaller corollas and glabrous anthers; differs from *P. mutelii* in that its inflorescence is not branched and the corolla is more inflated; differs from *P. schultzii* in the shorter calyx-teeth, and corolla straight or slightly curved (not geniculate).

Phelipanche daninii
Israel (left, right)

NAME ***Phelipanche georgii-reuteri*** Carlón, G.Gómez, M.Laínz, Moreno Mor., Ó.Sánchez & Schneew.

Habit: slender; inflorescence dense, shorter than its stem. **Height**: 10–27 cm. **Stem**: simple, rarely branched at the base, grey-violet, glandular-pubescent. **Calyx**: glandular-pubescent; teeth triangular with subulate apex, grey-lilac, shorter or mostly the same length as the tube; teeth 3–4 mm long. Lower flowers sessile. **Corolla**: erecto-patent, tubular-infundibuliform; dorsal line of the corolla with almost constant curvature strongly flexed forwards near the upper lip; 14–20 mm; violet, with whitish mouth; lanuginose. Upper lip bilobed, porrect, ± ovate and with entire margin; lower lip lobes ovate, entire, almost equal size, deflexed. **Carpel**: stigma lobes white. **Stamens**: filaments inserted 2.5–5 mm above corolla base, sparsely glandular-pubescent at the base, glabrous above; anthers almost glabrous. **Host**: parasitic on Brassicaceae, usually on *Lepidium subulatum.* **Ecology and distribution**: semi-arid grasslands and meadows, roadsides and forest clearings. Flowering May–July. Endemic to central and southern Spain. **Similar species**: differs from *P. mutelii* mainly by the erecto-patent corolla that is strongly flexed forwards near the upper lip.

Phelipanche georgii-reuteri
Spain (right)
photo: Gonzalo Moreno Moral

NAME ***Phelipanche gratiosa*** (Webb & Berthel.) Carlón, G.Gómez, M.Laínz, Moreno Mor., Ó.Sánchez & Schneew.

Habit: slender; inflorescence dense, shorter or of the same length as its aerial stem. **Height**: 15–35 cm. **Stem**: grey, dull violet or brownish, simple, rarely branched, glandular-pubescent. **Calyx**: yellowish, glandular-pubescent; teeth lanceolate-subulate, dark yellow, equalling to longer than their tube; teeth 5–7 mm long. Floral bracts yellowish, 7–9 mm, triangular-lanceolate to ovate-lanceolate, acute, shorter or equalling calyx. Bracteoles linear-subulate. Lower flowers not pedicellate or subsessile. **Corolla**: erect, tubular-infundibuliform with long tube and dorsal line slightly curved or almost straight; 19–27 mm; pale violet to blue-violet distally, with darker veins, lilac and whitish in the ventral part. Upper lip bilobed; folded, ± ovate and with entire margins; lower lip lobes ovate, elongated, with entire margins, almost equal size, deflexed. **Carpel**: stigma lobes whitish. **Stamens**: filaments inserted 4–7 mm above corolla base, sparsely and conspicuously non-glandular-pubescent at the base; anthers glabrous. **Host**: parasitic on *Launaea arborescens* (Asteraceae) or rarely other Asteraceae (e.g. *Reichardia tingitana* and *Urospermum picroides*). **Ecology and distribution**: ruderal habitats including roadsides, volcanic slopes, fixed dunes and hillsides. Flowering February–March. Endemic to the Canary Islands (Fuerteventura, Graciosa, Gran Canaria and Lanzarote); the commonest species on the eastern islands. **Similar species**: differs from *P. mutelii* by its larger bracts, erect flowers, narrower corolla tube, host, distribution and ecology.

Phelipanche gratiosa UK (above) cross section (cultivated plant), Fuerteventura, Canary Islands (right)

NAME ***Phelipanche gussoneana*** **(Lojac.) Domina, Raab-Straube, Rätzel & Uhlich**

Habit: slender; inflorescence dense, shorter than or of the same length as its stem. **Height**: 10–30 cm. **Stem**: simple, rarely branched, lilac, glandular-pubescent. **Calyx**: slightly glandular-pubescent; teeth lanceolate-subulate, yellowish-grey, mostly the same length as the tube; teeth 8–10 mm long. Floral bracts greyish, 4–7 mm, lanceolate, acute, shorter than calyx. Bracteoles linear-subulate. Lower flowers usually with short pedicels. **Corolla**: erect-spreading, straight or deflexed near the upper lip; 16–23 mm; pale violet to blue-violet distally, with darker veins. Upper lip bilobed, folded, ± ovate and with entire margins; lower lip lobes rounded, with entire margins, almost equal size, slightly deflexed. **Carpel**: stigma lobes whitish. **Stamens**: filaments inserted 2.5–4 mm above corolla base, glabrous or rarely slightly hairy; anthers hairy. **Host**: a generalist, parasitic on Asteraceae, Brassicaceae, Geraniaceae, Polygonaceae and other families. **Ecology and distribution**: fertile pastures, grassland, scree. Flowering April–June. Native to France, Sicily, Balkans, Caucasus and Transcaucasus. **Similar species**: differs from *P. mutelii* mainly by its darker stem, darker bluish veins on the corolla and the straight dorsal line of the corolla; differs from *P. oxyloba* by its larger flowers and longer, filiform, calyx-teeth.

Phelipanche gussoneana
Sicily, Italy (right)

NAME ***Phelipanche hajastanica*** Piwow., Ó.Sánchez & Moreno Mor.

Habit: usually robust; inflorescence (3–)5–6(–8) cm, ovate, rarely short-cylindrical, shorter than its stem and frequently ± comose at apex, ± dense. **Height**: (9–)12–14(–20) cm. **Stem**: simple, unbranched, usually robust, very densely glandular-pubescent, but white-lanate or villous in middle and upper parts, usually dark violet, base rarely yellowish; scales (7–)10–12(–14) mm, ovate-lanceolate to lanceolate, densely white glandular-pubescent. **Calyx**: segments narrowly triangular-lanceolate with subulate apices, veined, longer than the calyx tube; (9–)11(–13) mm long, shorter than the width of the corolla. Floral bracts (9–)10–11(–12) mm long, triangular lanceolate to lanceolate, shorter than calyx, with dense white glandular hairs. Bracteoles shorter than calyx, linear-lanceolate. **Corolla**: erecto-patent, tubular-infundibuliform, with slightly curved dorsal line, base slightly inflated, becoming narrow above the stamen insertion point; (15–)19–20(–22) mm long, (4.0–)4.5–5.0(–6.0) mm in diameter in central part; deep violet (rarely pale), lobes violet to blue-violet, with violet veins and sinuous edges; very densely glandular-pubescent externally, as well as on the inner lips. Upper lip bilobed, slightly emarginate, lobes sometimes with subacute apex, densely glandular-pubescent; lower lip lobes of similar size, ovate, densely ciliate with glandular hairs; folds white, very wide, with very dense white glandular hairs. **Carpel**: stigma quadrangular to discoid, slightly bilobed with 2 ± spherical white-blue or yellowish lobes; style with sparse glandular hairs. **Stamens**: filaments obliquely inserted at 4–5(–6) mm above corolla base, glabrous or with scarce non-glandular hairs at the base or near the anthers; anthers glabrous or rarely basally along suture with scarce long hairs. **Host**: *Nepeta trautvetteri* (Lamiaceae). **Ecology and distribution**: rocky and sandy slopes with sparse vegetation, at 1,400 m. Flowering end May–June, fruiting June–July. Known from the Lesser Caucasus, in southern Armenia. **Key diagnostics**: unbranched, dark violet, white-lanate or villous inflorescences with acute and ± comose apex; flowers erecto-patent and white densely glandular-pubescent; floral bracts shorter than calyx; calyx-teeth longer than their tube; violet corolla up to 22 mm; only on *Nepeta*. **Similar species**: similar to other lanate species such as *P. pulchella*, *P. zangezuri* and *P. caesia*, but differs in the characteristics stressed above, and in its host.

Phelipanche hajastanica
Armenia (left, right)

NAME ***Phelipanche heldreichii*** (Reut.) Soják
SYNONYMS ***Orobanche heldreichii*** (Reut.) Beck; ***Orobanche hohenackeri*** (Reut.) Tzvelev

Habit: slightly robust to slender; inflorescence usually longer than, or as long as, its stem, frequently with many scales, in a compacted, long and typical comosus apex. **Height**: (14–)21–23(–33) cm. **Stem**: stout or slender, with 2–8 smaller branches growing from the base; violet to blue, rarely yellowish, contrasting scales, floral bracts and calyx; scales (11–)14–16(–19) mm ovate to lanceolate. **Calyx**: teeth lanceolate with filiform tips, 1–2 as long as tube; teeth (7–)10–12(–15) mm, tube (3.0–)4.0–4.5(–6.0) mm. Floral bracts (2.5–)3.0–3.5(–3.8) mm, wide narrowly lanceolate to linear-lanceolate, slightly shorter than or as long as the calyx. **Corolla**: tubular, slightly infundibuliform in the distal part with dorsal line ± straight; (16–)18–21(–22.5) mm; violet to blue, glandular-pubescent. Upper lip 2-lobed, lobes acute, rarely ovate, ciliate; lower lip lobes smaller and narrower, of similar size to one another, ovate, usually blunt or sometimes acute, rarely apiculate, ciliate. **Carpel**: stigma scarcely bilobed, white or whitish-yellowish; style usually glandular-pubescent. **Stamens**: filaments sparsely pubescent at the base or sub-glabrous; anthers usually glabrous or with sparse shorter hairs, sometimes denser. **Host**: exclusively on *Eryngium* spp. (Apiaceae), especially *E. campestre* and *E. billardierei*. **Ecology and distribution**: rocky, grassy, loamy slopes, hills, steppe, semi-deserts, roadsides and hills, usually at (1,400–)1,600–2,000(–2,500) m. Flowering (end May) June–July, fruiting July–August. Turkey (Anatolia), Azerbaijan, Armenia and Iran; probably across south-western Asia. **Key diagnostics**: simple to much-branched, violet to blue (rarely yellowish) stems contrasting with the scales, bracts and calyx; inflorescence long; floral bracts slightly shorter than or as long as the calyx; corolla tubular (slightly infundibuliform in distal part), ± uniformly violet to blue; anthers usually glabrous; host-specific. **Similar species**: similar to other species with long, subulate calyx-teeth, such as *P. schultzii* and *P. libanotica*. In the Caucasus, *P. heldreichii* is morphologically closest to *P. sevanensis*, which is a larger plant, with an inflorescence shorter than its stem, branches if present, arising from the central part of the spike, larger corolla widened at the mouth with curved dorsal line, large, broad lobes, larger calyx tube with shorter calyx-teeth, and which has a different bract morphology, colouration, host and flowering time.

Phelipanche heldreichii
Armenia (left, right)

NAME ***Phelipanche hirtiflora*** (Reut.) Soják
SYNONYMS ***Orobanche hirtiflora*** (Reut.) Burkill; ***Phelipanche astragali*** (Mouterde) Holub; ***Phelipanche cohenii*** (Domina & Danin) Domina

Habit: usually robust, clumped with variably sized stems. **Height**: 15–20 cm. **Stem**: simple, rarely branched (or branched from the base), robust, glandular-pubescent, violet to pale brown. **Calyx**: densely glandular-pubescent with triangular-lanceolate, light yellow segments, shorter than to equalling their tube; calyx-teeth 3–4.5 mm long. Floral bracts 10–15 mm, ovate-lanceolate, shorter than to equalling the calyx. Bracteoles linear-subulate, 5–7 mm long. Lower flowers sessile. **Corolla**: straight, infundibuliform in the distal part and constricted above the ovary, inflated near the upper lip; 19–25(–30) mm, yellowish-violet. Upper lip bilobed, folded, ± rounded with entire margins; lower lip lobes ovate, elongated, entire, almost equal in size, deflexed. **Carpel**: stigma lobes yellowish. **Stamens**: filaments inserted 4–7 mm above corolla base, hairy; anthers glabrous. **Host**: parasitic on *Astragalus* spp. (Fabaceae). **Ecology and distribution**: on rocky-stony soil, in tragacanth-dominated scrub. Flowering period April–June. Occurring from the eastern Mediterranean and Caucasus to western Pakistan. **Similar species**: differs from *P. schultzii* in the shorter calyx-teeth, corolla straight or slightly curved (not geniculate); and from *P. aegyptiaca* mainly in that the flowers are not pedunculate, and in the smaller corollas and glabrous anthers.

Phelipanche hirtiflora
Israel (right)
photo: Avinoam Danin

NAME ***Phelipanche lavandulacea*** (Rchb.) Pomel
SYNONYMS ***Orobanche lavandulacea*** Rchb.; ***Phelypaea lavandulacea*** (Rchb.) Reut.; ***Kopsia lavandulacea*** (Rchb.) Caruel

Habit: robust; inflorescence (12–)19–47 cm, cylindrical, dense, many-flowered. **Height**: (20–)25–45(–50) cm. **Stem**: often with many (up to 14) branches; usually dark purplish or violet, but often yellowing at spike axis, glandular-pubescent; scales (8–)12–15 mm, broadly lanceolate, acuminate. **Calyx**: campanulate with teeth generally shorter or of similar length to the tube and subulate, broadly lanceolate, gradually narrowing, with glandular hairs; 9–11 mm. Bracts 9–13 mm, broadly lanceolate, dark blue. Bracteoles usually linear-lanceolate. Lower zone of the inflorescence with pedicellate flowers. **Corolla**: ± narrowly tubular, slightly widened towards the throat, rather narrow proximally, deflexed towards ⅓ of its length, sub-campanulate, somewhat curved; 16–22 mm; lilac to dark greyish blue, violet, white towards the ventral side, with 2 whitish folds on the throat, glandular-pubescent. Upper lip bilobed, with lobes ± broadly rhombic-ovoid or obovoid, ± patent; lower lip lobes obtuse, suborbicular, the folds white and hairy. **Carpel**: stigma lobes white, rarely yellowish; style sub-glabrous. **Stamens**: filaments inserted 3–5 mm above the base of the corolla; with sparse glandular hairs towards the apex, ± glabrous towards the base; anthers hairy. **Host**: only on *Bituminaria bituminosa* (Fabaceae). **Ecology and distribution**: widespread in xerothermic grasslands, rocky slopes, ruderal sites and on roadsides. Flowering period (end April) May–June (July). Mediterranean Basin west to the Canary Islands, Balkans, east to Russia, and across the Middle East. **Key diagnostics**: a tall, often branched. many-flowered inflorescence with pedicellate flowers below; corolla narrowly tubular, usually deep violet, lower lip of the corolla obtuse; calyx-teeth of similar length to tube; host-specific. **Similar species**: *Phelipanche lavandulaceoides* is very similar, and on the same host, but differs in its sub-funnel-shaped corolla with lobes broadly spathulate, calyx with longer teeth than the tube and gradually (not abruptly) attenuated, and usually simple stem; the entire plant is more intensely violet. *Phelipanche schultzii* differs in having an inflorescence with few or no pedicellate flowers, acute corolla lower lip and calyx-teeth considerably longer than their tube, as well as a different host range (Apiaceae).

Phelipanche lavandulacea
Kos, Greece (left)
Montenegro (right, page 536, 537)

NAME ***Phelipanche lavandulaceoides*** Carlón, G.Gómez, M.Laínz, Moreno Mor., Ó.Sánchez & Schneew.

Habit: robust; inflorescence cylindrical, rather lax, many-flowered (24–49 flowers). **Height**: 19–59 cm. **Stem**: often with many (up to 14) branches; usually intensely dark purplish or violet, glandular-pubescent; scales 8–16 × 3–5 mm, triangular-lanceolate, intensely violet, blackish when dry. **Calyx**: campanulate with teeth generally shorter than or of similar length to the tube and subulate, broadly lanceolate, gradually narrowing, with glandular hairs; 9–11 mm. Bracts 8–14 × 2.6–4.5 mm, lanceolate-triangular, as long as the calyx or a little longer; intense violet and covered on its external face by hairs similar to those of the stem. **Corolla**: patent, sub-infundibuliform (4.42–5.12 mm wide at the throat and 1.94–2.30 mm wide at the constriction); with multicellular hairs topped with yellow glands, longer and abundant on the back; 19–22 mm long; intensely violet except at the end of the ventral side, where it is whitish; throat ornamented by 2 contrasting white folds, very densely hairy long, multicellular, hairs, not glanduliferous, white (translucent when dry); lobes large, ± spathulate, rounded at the apex and very hairy, hairs long, multicellular, non-glanduliferous, violet (dry, translucent, with the septa violet or brown). **Carpel**: disciform-bilobed and ± creamy or yellowish; style whitish to violet, glabrous or, towards the apex and base, with some glandular hairs. **Stamens**: filaments with a somewhat dilated base, inserted 3.2–6.2 mm (adaxial) and 2.7–5.5 mm (abaxial) from the base of the corolla, slightly hairy at the base and beneath the anthers. **Host**: only on *Bituminaria bituminosa* (Fabaceae). **Ecology and distribution**: endemic to inner Spain and particularly the Ebro basin. **Key diagnostics**: sub-funnel-shaped corolla with broadly spathulate lobes, calyx with teeth longer than their tube, simple stem or with few branches; intense violet colouration; host-specific. **Similar species**: the level of distinction from *P. lavandulacea* is unclear, *P. lavandulaceoides* reputedly differs in having sub-funnel-shaped corolla with lobes broadly spathulate, calyx with longer teeth than the tube and gradually attenuated, and (usually) simple stem.

Phelipanche lavandulaceoides
Spain (right)
photo: Gonzalo Moreno Moral

NAME ***Phelipanche libanotica*** (Schweinf. ex Boiss.) Soják
SYNONYMS ***Orobanche libanotica*** (Schweinf. ex Boiss.) Beck; ***Phelipanche orientalis*** (Beck) Soják; ***Phelipanche sogdiana*** (Novopokr.) Soják

Habit: somewhat robust; inflorescences dense and many-flowered, with acute and ± comose apex, usually longer than its stem. **Height**: up to 40(–60) cm. **Stem**: usually ± robust, branched (branches erect, clustered, almost parallel to the stem) or simple, brownish-yellow, sometimes with a purple tint, glandular-pubescent; scales 7–12(–20) mm, ovate to ovate-lanceolate, usually numerous and conspicuous. **Calyx**: teeth triangular-subulate, firm, usually clearly longer than calyx tube; (7–)9–11(–12) mm. Floral bracts (6–)8–10.5(–12) mm, ovate-lanceolate, often reflexed. Bracteoles linear-subulate. **Corolla**: tubular-infundibuliform to narrowly campanulate, 14–19 mm; various shades of blue, violet, whitish at the base, glandular-pubescent, sometimes entirely yellow. Upper lip deeply 2-lobed, lobes erect, triangular-lanceolate with subulate apex, ± hairy; lower lip deeply 3-lobed, lobes oblong-lanceolate, acute, nearly entire or finely crenate, folds and lobes hairy. **Carpel**: stigma white, rarely yellow-white; style sub-glabrous. **Stamens**: filaments glabrous or sub-glabrous; anthers ± densely hairy. **Host**: exclusively on wild *Prunus* spp. (syn. *Amygdalus* spp.), e.g. *P. fenzliana* and *P. bucharica* (Rosaceae), and cultivated apricots *P. armeniaca* in Turkey, where it is a noxious weed. Also reports in vineyards, on *Vitis* (Vitaceae). **Ecology and distribution**: rocky slopes, semi-deserts, steppe, fields. Flowering usually May–June (July), fruiting (June) July–August. Distribution: north-west Himalayas (India and Pakistan), Afghanistan, Iran, Iraq, Jordan, Lebanon, Syria, Tajikistan, Kyrgyzstan, Uzbekistan, Turkey, the Caucasus (Armenia, Georgia and Azerbaijan) and south-east Europe (Cyprus and Greece). Probably widespread across western and central Asia. **Key diagnostics**: usually robust with parallel branches; inflorescences dense and many-flowered, with characteristic acute and ± comose apex; corolla deeply lobed, lobes acute, with subulate apex; on tree roots, usually *Prunus*. **Similar species**: confused with *P. schultzii*, which is taller (up to 65 cm) and differs in its floral morphology (e.g. calyx with linear-subulate or filiform teeth, 2–3 times longer than their tube), and host range.

Phelipanche libanotica
Armenia (right, page 542, 543)

NAME ***Phelipanche mutelii*** (F.W.Schultz) Pomel
SYNONYMS ***Kopsia mutelii*** (F.W.Schultz) Bég.; ***Phelypaea mutelii*** (F.W.Schultz) Reut.; ***Orobanche ramosa*** **subsp.** ***mutelii*** (F.W.Schultz) Cout.

Habit: slender; inflorescences: ± dense above, sometimes lax below, shorter than or equalling its stem. **Height**: 10–25 cm. **Stem**: simple or branched (with several small, erect branches), glandular-pubescent, yellowish, violet-tinted; scales brown, few or absent above-ground. **Calyx**: teeth lanceolate-subulate, yellowish to violet, equalling to longer than the tube; teeth 6–11 mm long, glandular-pubescent. Floral bracts 4–6 mm, ovate-lanceolate, obtuse, shorter than calyx. Bracteoles linear-subulate. Lower flowers usually not pedicellate, rarely with short pedicel. **Corolla**: patent or erecto-patent, straight or slightly curved, infundibuliform in the distal part constricted above the ovary, slightly inflated near the upper lip; 15–22 mm; pale violet to blue-violet distally, with darker veins, whitish in the ventral part. Upper lip bilobed, folded, ± ovate and entire; lower lip lobes ovate, elongated, entire, almost equal size, deflexed. **Carpel**: stigma 2-lobed, rarely 4-lobed, lobes whitish. **Stamens**: filaments inserted 3–5 mm above corolla base; hairy below, glabrous or rarely slightly hairy above; anthers glabrous. **Host**: Asteraceae (*Calendula*, *Crepis*, *Hedypnois*, *Lactuca*, *Leontodon*, *Reichardia* and *Sonchus*), rarely on other families. **Ecology and distribution**: scrub, field margins, semi-ruderal pioneer xerothermic communities, roadsides, meadows, xerothermic grasslands. Flowering April–May. Native to the Mediterranean region, need confirmation reports from Caucasus, Transcaucasus, Iraq, Iran and Yemen. **Similar species**: differs from *P. ramosa* mainly by its larger, patent, flowers with lower lip deflexed; differs from *P. aegyptiaca* by the lack of pedunculate flowers, smaller corollas and glabrous anthers; differs from *P. daninii* by its branched inflorescence and less inflated corolla; differs from *P. nana* in its distinctly gibbous corolla profile, oblong-lanceolate lower lip, ± entire and acute, and triangular-lanceolate calyx-teeth with subulate to filiform apex, slightly exceeding their tube; *P. hedypnoidis* Rätzel, Ristow & Uhlich differs in having a calyx with 5 or more teeth, and in corolla colouration (yellowish, dorsally and terminally tinged blue violet).

Phelipanche mutelii Sicily, Italy (above), Crete, Greece (right)

NAME ***Phelipanche nana*** (Noë ex Rchb.) Soják
SYNONYMS ***Orobanche melitensis*** Beck; ***Orobanche nana*** Noë ex Rchb.; ***Orobanche ramosa* subsp. *nana*** (Noë ex Rchb.) Rouy; ***Phelipanche mutelii* var. *nana*** (Noë ex Rchb.) Uhlich & Rätzel

Habit: small, slender; inflorescence ovoid or shortly cylindrical, usually dense but few-flowered, usually equalling or shorter than its stem. **Height**: 6–25 cm (rarely more). **Stem**: simple or branched, yellowish, yellowish-brown, glandular-pubescent; scales 5–8 mm, ovate-lanceolate. **Calyx**: teeth triangular-lanceolate with subulate to filiform apex, 5–8 mm long, equalling or slightly longer than their tube. Floral bracts up to 8 mm long, ovate-lanceolate. Bracteoles linear-subulate. Flowers patent, rarely erecto-patent. **Corolla**: irregularly tubular-infundibuliform (dorsal and ventral lines unequal), geniculate. Corolla cross section ± broadly elliptical or ovoid. Dorsal line of the corolla slightly bulging above the geniculate. Corolla upper outline distinctly gibbous; 11–15(–17) mm; blue to violet (deep or less frequently pale), rarely whitish or yellowish, glandular-pubescent outside. Upper lip deeply bilobed, folded, ± ovate and entire, divergent; lower lip lobes not especially divergent, oblong-lanceolate, broadly ovate or obtuse, ± entire and acute, folds hairy. **Carpel**: stigma bilobed, white, rarely pale blue, pale yellow; style sparsely glandular-pubescent below stigma. **Stamens**: filaments sparsely pubescent at base, sub-glabrous; anthers usually glabrous. **Host**: various families, including Asteraceae, Geraniaceae, Oxalidaceae, Apiaceae, Rubiaceae, Plantaginaceae, Brassicaceae; usually on wild plants, e.g. *Caucalis platycarpos*, *Orlaya grandiflora*, *Erodium cicutarium*, *Geranium robertianum*, *Sherardia arvensis* and *Plantago lanceolata*. **Ecology and distribution**: rocky slopes, xerothermic grasslands, steppe, meadows, pastures, ruderal habitats, roadsides, lawns, urban spaces, forests and shrubland margins. Flowering (April) May–June, fruiting June to July. Throughout the Old World (Mediterranean Basin, Balkans, Asia Minor and the Caucasus). Reportedly introduced to the Americas and Australia; these populations may correspond better with the closely related *P. ramosa*. **Key diagnostics**: polymorphic, usually short with patent flowers, corolla dorsal line gibbous, lower lip lobes oblong-lanceolate, ± entire and acute, calyx-teeth slightly longer than tube, triangular-lanceolate with subulate to filiform apex; parasitic on wild plants. **Similar species**: distinguishing *P. nana* from *P. mutelii* and *P. ramosa* is challenging: *P. mutelii* has usually erect to erecto-patent tubular-infundibuliform corolla of regular shape, with an open throat and dorsal line ± straight above the staminal constriction; corolla dorsal line not or scarcely gibbous, lobes of upper lip flat, ± orbicular and erose; lobes of lower lip ± divergent, broadly rhomboidal or oblong-spathulate, ± erose and obtuse (rarely apiculate). *Phelipanche nana* is often treated as a subspecies of *P. ramosa*. The latter differs in its typically branched stem and longer, lax, inflorescence, flowers more erect, paler blue to lilac with rounded lower lip lobes and acuminate calyx-teeth equalling or shorter than their tube; typically later flowering, and parasitic on crops.

Remarks: *Phelipanche oxyloba* (Reut.) Soják differs from *P. nana* subtly in that the corolla is not widened at the throat; parasitic on Asteraceae, in southern and eastern Europe (Greece and Turkey) to central Asia and west Pakistan.

Phelipanche nana
Sicily, Italy (right)

NAME ***Phelipanche nowackiana*** (Markgr.) Soják
SYNONYMS ***Orobanche nowackiana*** Markgr.; ***Orobanche caucasica* f. *flava*** Gilli; ***Orobanche caucasica* f. *fuscolutea*** Gilli; ***Phelipanche tzvelevii*** Teryokhin

Habit: small, slightly robust. **Height**: 5–15 cm. **Stem**: unbranched, slender, arachnoid-glandular-villous in the upper part; yellowish-brownish. Inflorescence shortly cylindrical, dense, few-flowered, white-villous. **Calyx**: campanulate, glandular, densely villous, 4-dentate; tube 1–5 mm, with 3 larger teeth 6–10 mm, distinctly longer than their tube, several-veined, other tooth much smaller, 2 mm, bifid. Floral bracts lanceolate, subequalling or shorter than the calyx, densely white-villous. Bracteoles 2, linear, longer than the calyx tube. **Corolla**: 17–25 mm, constricted at the point of filament insertion, inflated below insertion, gradually dilated above the insertion to the limb, glandular and arachnoid-lanate, erecto-patent or forward-inclined; pale yellow, becoming brownish-yellow when dry. **Carpel**: stigma pale yellow, pilose; style sparsely glandular above. **Stamens**: filaments included; anthers densely arachnoid-villous towards the base, cream to pale yellowish brown. **Host**: Brassicaceae: *Odontarrhena* spp. and *Bornmuellera* spp.. **Ecology and distribution**: serpentine rock, in open *Pinus nigra* forest and on rocky limestone slopes, at elevations of 1,500–2,000 m; from Albania through Greece, North Macedonia, Serbia to Turkey. This parasite and its hosts associated with serpentine soils are hyperaccumulators of nickel. Flowering May–June. **Key diagnostics**: very short dense inflorescences; yellow corolla; broad scales and bracts. **Similar species**: difficult to confuse owing to its unique yellow colour, and the characteristics stressed above, but morphologically most similar to the *Phelipanche pulchella* group (*P. bungeana*, *P. caucasica* and *P. tzvelevii*).

Phelipanche nowackiana Greece (above) photo: Todd Boland, Kosovo (right, page 550, 551) photos: Besnik Fetiu

NAME ***Phelipanche olbiensis*** (Coss.) Carlón, G.Gómez, M.Laínz, Moreno Mor., Ó.Sánchez & Schneew.

SYNONYMS ***Phelypaea mutelii* subsp. *olbiensis*** (Coss.) Nyman; ***Phelypaea ramosa* subsp. *olbiensis*** (Coss.) Bonnier; ***Phelypaea ramosa* subsp. *mutelii* var. *olbiensis*** (Coss.) Rouy

Habit: small, slender, lax, with few flowers; inflorescence shorter than its stem. **Height**: 10–20 cm. **Stem**: simple, yellowish to violet, glandular-pubescent; scales 4–6 mm, triangular to ovate, obtuse, shorter than calyx. Bracteoles linear-subulate. **Calyx**: glandular-pubescent; teeth triangular with subulate apex, yellowish-brown, mostly the same length as their tube; teeth 4–5 mm long; lower flowers sessile. **Corolla**: erecto-patent, tubular to tubular-campanulate; dorsal line of the corolla is geniculate in the lower part and then ± straight; 15–18 mm; violet, basally yellowish-white, veined, with usually whitish mouth; sparsely, short glandular-pubescent. Upper lip bilobed; porrect, ± ovate and entire; lower lip lobes rounded, entire, almost equal size or the central longer than the lateral ones, deflexed. **Carpel**: stigma 2-lobed; lobes whitish. **Stamens**: filaments inserted 3–4 mm above corolla base; sparsely glandular-pubescent; anthers sparsely glandular-pubescent. **Host**: on *Helichrysum* spp. (Asteraceae), mainly *H. stoechas*; also reported on *Phagnalon saxatile* and *P. rupestre* (Asteraceae). **Ecology and distribution**: coastal dunes and cliffs, bare habitats and forest clearings. Flowering April–May. Native to the Mediterranean, known from Spain, Greece, Crete, Cyprus and Turkey. **Similar species**: differs from *P. rosmarina* mainly in its tubular-campanulate corolla shape, and in the bract and the teeth of the calyx being triangular, not lanceolate.

Phelipanche olbiensis
Spain (right)
photo: Gonzalo Moreno Moral

NAME ***Phelipanche portoilicitana*** (A.Pujadas & M.B.Crespo) Carlón, G.Gómez, M.Laínz, Moreno Mor., Ó.Sánchez & Schneew.

SYNONYM ***Orobanche portoilicitana*** A.Pujadas & M.B.Crespo

Habit: stout or sometimes slender; inflorescence (4–)7–15 cm, very dense, many-flowered, cylindrical, usually attenuate in the upper part and rounded at the apex. **Height**: (11–)15–34 cm. **Stem**: simple or branched, frequently with numerous scales, whitish, yellowish, upwards glandular-pubescent; scales 8–14 mm, triangular to ovate-lanceolate and lanceolate, shortly pubescent and densely arranged. **Calyx**: tube equalling segments in length; calyx-teeth lanceolate-subulate, covered with short glandular hairs, (7–)9–11 mm. Floral bracts 6–9 mm, broadly lanceolate, shorter than calyx. Bracteoles 6–8 mm, linear, equalling the bracts. **Corolla**: flowers sessile, usually erect, more rarely erecto-patent; tubular or slightly infundibuliform, somewhat galeate at the back, not swollen downwards and very slightly contracted in the insertion of the filaments; 16–21 mm long; pale blue, bluish-violaceous or bluish-rosaceous, whitish towards the base, densely clothed with short glandular hairs in the outer side and margins of lips ciliate. Upper lip shortly bilobed and somewhat emarginate, with lobes straight forward or patent; lower lip with subequal lobes, straight or deflexed, ovate, sub-obtuse and shortly apiculate, and with dentate margins. **Carpel**: stigma usually white, rarely yellowish; style sparsely pilose-glandulose, with short glandular hairs more densely at the apex. **Stamens**: filaments inserted 4–6 mm above the base of the corolla tube; glabrous; anthers with several hairs at the base, sometimes glabrous. **Host**: *Centaurea* spp. (Asteraceae), especially *C. seridis*, *C. aspera* and *C. behen*. **Ecology and distribution**: deep sandy substrates, coastal sand dunes, loamy slopes and garrigue. Flowering mid-June–early July, fruiting July–August. Mainly Mediterranean Basin: Algeria, Egypt, Libya, Morocco, Spain and Tunisia, and recent discoveries from western Asia (Armenia). **Key diagnostics**: thick stems with numerous bracts, dense, many-flowered inflorescence with erect, pale blue flowers; lobes of lower lip ovate, sub-obtuse and shortly apiculate; filaments glabrous. *Centaurea*-specific. **Similar species**: *Phelipanche heldreichii* has a similar ecology in Asia but lacks the characters stressed above and grows on *Eryngium*. Other species with calyx-teeth longer than or equalling their tube include *P. schultzii*, which is generally much more robust (up to 65 cm) and differs in floral morphology (e.g. linear-subulate or filiform calyx-teeth 2–3 times longer than their tube, acute lower lip corolla lobes), and in host range.

Phelipanche portoilicitana
Armenia (right, page 556)
Spain (page 557)
photo: Gonzalo Moreno Moral

NAME ***Phelipanche pulchella*** (C.A.Mey.) Soják
SYNONYMS ***Orobanche pulchella*** (C.A.Mey.) Novopokr.; ***Orobanche caucasica*** Beck; ***Phelipanche bungeana*** (Beck) Soják; ***Phelipanche septemloba*** (Beck) Soják

Habit: slender; inflorescence usually ovoid, sub-capitate, much shorter than its stem, 3–7(–10) cm, lax, few-flowered (or rarely up to 20 flowers); flowers large relative to plant. **Height**: (10–)15–20(–25) cm. **Stem**: simple or rarely branched (at most twice), slender, yellowish-brown, violet-brown, usually glandular-hairy or arachnoid-villous; scales (8–)10–13(–15) mm, lax, lanceolate to ovate-lanceolate. **Calyx**: segments usually shorter than or equalling the corolla tube, narrowly lanceolate, often subulate at the tips, equalling or slightly longer than the calyx tube; usually greyish with dense glandular pubescence and lanate, with white hairs; (8–)10–15(–18) mm. Floral bracts (10–)11–12(–13) mm, ovate-lanceolate, shorter than or almost equalling calyx, usually arachnoid-villous, lanate, with white hairs. Bracteoles linear-subulate, slightly shorter than calyx, arachnoid-villous. **Corolla**: ± erect or erecto-patent, usually lanate, with characteristic longer stiff hair; infundibuliform, rarely narrowly tubular, gradually broadened towards the limb, more erecto-patent, dorsal line straight or slightly curved; (17–)22–25(–27) mm; bright blue to blue-violet, lighter at the base of the tube and with darker veins, usually with dense and rather long indumentum on the exterior and glabrous inside except for the densely hairy folds and sparsely hairy limb lobes. Upper lip 2-lobed, lobes ciliate and crenate along the margins; lower lip with middle lobe equal to or slightly larger than the laterals; lobes orbicular or broadly ovate, rarely subacute. **Carpel**: stigma white, rarely whitish-bluish; style sparsely glandular-pubescent near the stigma or glabrous. **Stamens**: filaments insertion 4–5 mm above corolla tube base, glabrous or with sparse hairs at the base; anthers hairy at base and along fusion. **Host**: *Teucrium chamaedrys*, occasionally *T. polium* or *Scutellaria orientalis* (Lamiaceae). **Ecology and distribution**: rocky and loamy slopes and subalpine grasslands, scrub, forest edges, ruderal habitats, slopes and roadsides. Flowering (April) May–June (July), fruiting (June) July–August. Caucasus (Armenia, Georgia and Azerbaijan) and surrounding areas (Iran and Turkey). **Key diagnostics**: small, slender ± white-woolly plant with short, sub-capitate, few-flowered inflorescence, flowers ± erect and large relative to the plant, corolla infundibuliform, gradually broadened towards the limb, lanate, with characteristic longer stiff hair, lanate calyx; host-specific. **Similar species**: *Phelipanche bungeana*, *P. caucasica* and *P. septemloba* are poorly known taxa that are frequently confused and recently considered to be synonyms of *P. pulchella*.

Phelipanche pulchella
Armenia (right, page 560, 561)

NAME ***Phelipanche purpurea*** (Jacq.) Soják
SYNONYMS ***Orobanche caerulea*** Vill.; ***Orobanche purpurea*** Jacq.; ***Kopsia purpurea*** (Jacq.) Bég.; ***Phelypaea purpurea*** (Jacq.) Asch.

Habit: usually slender; inflorescence usually lax, often many-flowered, elongated. **Height**: 15–60 cm. **Stem**: usually unbranched, slender, violet, blue-grey, rarely yellow, with short glandular hairs; scales sparse, lanceolate to narrowly lanceolate, up to 15(–20) mm long. **Calyx**: tubular-campanulate, 4–5 dentate, teeth narrow-lanceolate, equalling the tube or shorter, glandular-pubescent; 8–16 mm. Floral bracts lanceolate. Bracteoles are lanceolate, shorter than the calyx segments, with white glandular hairs. **Corolla**: erecto-patent to patent, ± tubular, narrowly campanulate; the dorsal line of corolla is almost evenly curved from the base onwards, sometimes bent upwards at the tip; 17–23(–25) mm; pale violet, blue, blue-violet, yellowish or whitish at the base, with dark violet veins, glandular-pubescent. Upper lip 2-lobed, with erect violet-veined lobes; lower lip lobes almost equal, narrowly ovate, acute, with ± entire edges or crenate, violet-veined and pubescent. **Carpel**: stigma bilobed, white, rarely yellowish or pale violet; style glandular-pubescent. **Stamens**: filaments inserted 5–7 mm above the base of the corolla tube; almost glabrous or sparsely glandular at the base; anthers glabrous (rarely sparsely hairy at sutures). **Host**: always on Asteraceae, mainly on *Achillea* spp., principally *A. millefolium*; sometimes also *A. collina*, *A. pannonica* and *A. arabica*, rarely on *Tanacetum partheniifolium*, *T. balsamitoides*, *Artemisia vulgaris* and *Tripleurospermum elongatum*. **Ecology and distribution**: temperate, arid and semi-arid grasslands, dry meadows, dunes and cliffs, rocky slopes, steppes, ruderal habitats, and roadsides. Flowering May–June (July), fruiting July–August. Most of Europe northwards to the UK, Baltic states and Russia and south to Asia Minor, Caucasus to eastern India. **Key diagnostics**: blue-grey colouration, sparse scales, erecto-patent to patent corolla; calyx-teeth equalling the tube; corolla ± tubular or narrowly campanulate, usually with darker longitudinal nerves; anthers usually glabrous; usually on *Achillea*. **Similar species**: similar to *P. arenaria*, which has yellowish-brown stems with numerous, triangular-lanceolate to lanceolate bracts, larger and erect to erecto-patent flowers, ± tubular-infundibuliform corolla, usually without darker longitudinal nerves, widely funnel-shaped in the upper part; woolly anthers; calyx segments longer than or sometimes equalling the tube; on *Artemisia*. *Phelipanche bohemica* is variably considered to be distinct from *P. purpurea*; its stem is usually stouter, up to 70 cm; it is more intensely coloured in all parts, more dark violet; the inflorescence is usually longer, dense, many-flowered, flowers slightly longer, corolla with a curved dorsal line, with ovate, obtuse lobes with somewhat eroded edges, and parasitic exclusively on *Artemisia campestris*.

Infraspecific taxa (selected)

Subsp. *purpurea* is the typical tall, lax-flowered-form with patent flowers, lanceolate lower lip lobes with white folds. Parasitic on *Achillea* and *Artemisia,* often in disturbed or ruderal places across western and central Europe and south-west Asia.

Subsp. *millefolii* is shorter, with a uniformly coloured corolla and rounded, apiculate lower lip lobes. Parasitic on *Achillea* in undisturbed habitats in western and central Europe.

Subsp. *ballii* is tall with rounded, not apiculate lower lip lobes. Parasitic on *Achillea ligustica* and *Cladanthus scariosus* in the Atlas Mountains (Morocco).

Phelipanche purpurea subsp. *purpurea* Georgia (right), Poland (page 564)
Phelipanche purpurea subsp. *millefolii* (page 565) photo: Gonzalo Moreno Moral

NAME **Phelipanche ramosa** (L.) Pomel

SYNONYMS ***Lathraea phelypea*** Forssk.; ***Orobanche albiflora*** Gren. & Godr., ***Orobanche ramosa*** L.; ***Kopsia ramosa*** (L.) Dumort.; ***Phelypaea ramosa*** (L.) C.A.Mey.

Habit: small, slender; inflorescence usually cylindrical, lax. **Height**: (8–)17–35(–45) cm. **Stem**: slender, erect, usually branched, yellow, yellowish, bluish or pale violet, glandular-pubescent; scales sparse, lax, lanceolate. **Calyx**: tubular-campanulate, usually 4-dentate; teeth equalling or shorter than the tube, triangular with an acute apex, glandular-pubescent; (3–)4–6 mm. Floral bracts are lanceolate, ± as long as corolla, glandular-pubescent. Bracteoles are broadly lanceolate, as long as or shorter than calyx segments, glandular-pubescent. **Corolla**: erect to erecto-patent, tubular, slightly constricted just above the ovary, then gradually opening out, pubescent. Corolla upper outline distinctly gibbous; 10–15(–17) mm; bluish, violet, whitish, rarely yellow. Upper lip 2-lobed, lobes porrect and pubescent; lower lip lobes ovate, orbicular, of about equal size, obtuse with hairy white folds, glandular-pubescent. **Carpel**: stigma bilobed, white, rarely pale blue or pale yellow; style sparsely glandular-pubescent. **Stamens**: filaments inserted 2–4 mm above the base of the corolla tube, sub-glabrous or sparsely glandular-pubescent at the base; anthers usually glabrous or sparsely hairy. **Host**: cultivated and ornamental plants, especially *Nicotiana tabacum* and *Lycopersicon esculentum*, *Solanum tuberosum* (Solanaceae), *Cannabis sativa* (Cannabaceae) and *Brassica napus* (Brassicaceae). **Ecology and distribution**: cultivated areas, mainly tobacco and tomato plantations, ruderal areas and gardens. Flowering usually late: end August–September, fruiting (August) September–October. Widespread species of the Old World (Europe, North Africa, Asia Minor, the Caucasus to central Asia) and naturalised in the Americas. **Key diagnostics**: a polymorphic parasite of cultivated plants: usually branched with small, erect, pale blue to lilac flowers; lobes of lower lip rounded; calyx-teeth equalling or shorter than their tube, acuminate, mainly late-flowering. **Similar species**: *Phelipanche nana* is very similar but typically shorter with patent (rarely erecto-patent), blue to violet (deep or less frequently pale) flowers; calyx-teeth slightly longer than the tube, triangular-lanceolate to filiform apex, and parasitic on wild plants; *P. mutelii* has a longer corolla, (15–)18–22 mm, and rounded lower lip lobes; *P. aegyptiaca* also infests crops but it has much larger, infundibuliform flowers (20–37 mm) and hairy anthers.

Phelipanche ramosa
UK (left, right)
Poland (page 568, 569)

NAME ***Phelipanche resedarum*** Carlón, G.Gómez, M.Laínz, Moreno Mor., Ó.Sánchez & Schneew.

SYNONYM ***Orobanche resedarum*** (Carlón, G.Gómez, M.Laínz, Moreno Mor., Ó.Sánchez & Schneew.) A.Pujadas & E.Triano

Habit: robust, congested, rather broad spikes with many flowers. **Height**: 14–29 cm. **Stem**: robust, simple to much-branched; branches arise from ground-level; stem creamy yellowish, sometimes tinged violet, with dense, long, translucent, multicellular, glanduliferous hairs with yellowish glands with a whitish appearance when dry. **Calyx**: 4, closely triangular teeth, reduced and narrow, sharp and ± equal to their tube; yellowish, and ± densely glandular-hairy; 8.5–15 × 5.3–6.5 mm. Floral bracts 6.5–12.3 × 2.0–4.3 mm, lanceolate-triangular to lanceolate, generally shorter than the calyx. **Corolla**: spreading (not erect); tubular-infundibuliform with a long tube, with a straight dorsal line, covered externally with relatively long glandular hairs; 18–24 mm wide, 2.5–4 mm at the constriction and 4.8–7.4 mm at the throat; ± pale blue, although more intense (violet) on the lobes, rarely, completely violet; throat with conspicuous whitish or cream folds. Upper lip bilobed, with erect, oblong-oval, hairy lobes; lower lip with deflexed, subequal, ± elliptical lobes. **Carpel**: stigma disciform, ± bilobed, yellow; style violet with some short glandular hairs. **Stamens**: filaments inserted 3–5 mm above the corolla tube base, generally covered with non-glanduliferous hairs at the base, whitish except towards the basal ⅓, where intensely yellow; anthers 1.45–1.67 × 1–1.32 mm, hairy, cream. **Host**: *Reseda* spp. (Resedaceae). **Ecology and distribution**: sunny slopes, rocky cliffs, hills, roadsides, thickets between crops, usually on chalky soil, in south-eastern Spain. Flowering May–June. **Key diagnostics**: dense spikes of spreading, long-tubed flowers with pale exterior and violet lip lobes with contrasting white folds; stigma lobes yellow; parasitic only on *Reseda.* **Similar species**: unlikely to be confused with other *Phelipanche* species on the basis of the characters stressed above.

Phelipanche resedarum
Spain (right)
photo: Gonzalo Moreno Moral

NAME ***Phelipanche reuteriana*** (Rchb.f.) Carlón, G.Gómez, M.Laínz, Moreno Mor., Ó.Sánchez & Schneew.

SYNONYMS ***Phelypaea reuteriana*** Rchb.f.; ***Orobanche reuteriana*** (Rchb.f.) M.B.Crespo & A.Pujadas

Habit: short, slender spikes with few flowers. **Height**: 7–13 cm. **Stem**: slender, white-woolly, brownish, pale yellow to dull red. **Calyx**: campanulate with short, lanceolate teeth shorter than their tube, white-woolly. Floral bracts ovate-lanceolate. **Corolla**: patent when mature, abruptly curved and inflated at the base then with a flat dorsal line, rather humped below the upper lip; 17–18 mm; interior and exterior corolla bright blue-violet, contrasting with the rather dull red, white-woolly stems; upper lip bilobed; lower lip with divergent, triangular-elliptic, subulate lobes with white folds. **Carpel**: stigma lobes contiguous, white. **Stamens**: filaments inserted at the point of constriction of the corolla tube, rather thickened below, and densely long-hairy; anthers cream. **Host**: *Plantago albicans* (Plantaginaceae); various other herbs suggested but requiring verification. **Ecology and distribution**: Spain and northern Africa; bare, sandy habitats. **Key diagnostics**: conspicuously white-woolly, few-flowered spikes; parasitic on *Plantago.* **Similar species**: unlikely to be confused with other *Phelipanche* species on the basis of the characters stressed above.

Phelipanche reuteriana
Spain (right)
photo: Gonzalo Moreno Moral

NAME ***Phelipanche rosmarina*** (Welw. ex Beck) Banfi, Galasso & Soldano
SYNONYMS ***Phelypaea rosmarini*** Welw.; ***Orobanche mariana*** A.Pujadas, ***Orobanche rosmarina*** Welw. ex Beck

Habit: slender, lax to rather congested; inflorescence longer than its stem. **Height**: 5–25 cm. **Stem**: simple, rarely branched, yellowish- to purplish grey, glandular-pubescent. **Calyx**: calyx-teeth triangular-lanceolate with subulate to filiform apex, yellowish-grey, mostly the same length as the tube; teeth 3–4 mm long, glandular-pubescent. Floral bracts 4–7 mm, ovate-lanceolate, obtuse, shorter than calyx. Bracteoles linear-subulate. Lower flowers: sessile. **Corolla**: erecto-patent, irregularly tubular-infundibuliform (dorsal and ventral lines unequal), geniculate, with a maximum subapical width of 15–18 mm, pale lilac to violet-whitish in the throat and internally. Upper lip bilobed, folded, ± ovate and entire; lower lip lobes rounded, entire, almost equal size, deflexed. **Carpel**: stigma 2-lobed; lobes whitish. **Stamens**: filaments inserted 2.5–3.5 mm above the corolla base; glabrous or rarely slightly hairy; anthers glabrous. **Host**: only on *Salvia rosmarinus* (Lamiaceae). **Ecology and distribution**: garrigue or maquis, and coastal cliffs where *S. rosmarinus* is abundant. Flowering March–June, depending on latitude: west and central Mediterranean from Portugal to Croatia, south to North Africa in Algeria and Tunisia. **Key diagnostics**: irregularly tubular-infundibuliform corolla profile; specificity for rosemary in garrigue-like habitats. **Similar species**: differs from *Phelipanche nana* in having a funnel-shaped, narrower rather than gibbous corolla; differs from *P. olbiensis* in corolla shape, and in having lanceolate, rather than triangular calyx segments.

Phelipanche rosmarina
Tunisia (left)
Spain (right) photo: Gonzalo Moreno Moral

NAME ***Phelipanche rumseyana*** A.Pujadas & P.Fraga

Habit: short, slender spikes with few flowers. **Height**: (7–)10–26(–35) cm. **Stem**: 2–4.5 mm in diameter at medium height, only slightly thickened or clavate at the base with a diameter of (5–)7–15 mm, simple, rarely branched below-ground, sometimes with sterile buds at the base, densely pubescent. **Calyx**: with 4 triangular teeth, long-acuminate, equalling the tube, lacking a fifth adaxial tooth, veins somewhat marked, calyx segments connate, abaxially up to 3.5 mm high, adaxially up to 1.5 mm high; segments purplish-blue, deeper purple-blue when dry. **Corolla**: erecto-patent to patent, dorsal line uniformly curved, the base slightly inflated, becoming narrow above the point of stamen insertion, 1.7–2.4 mm wide, slightly infundibuliform towards the apex, 3.5–5 mm wide at the mouth, shortly-pubescent; pale, dirty white outside; lobes blue-purple. **Carpel**: stigma scarcely bilobed, yellowish-white. **Stamens**: filaments inserted sub-horizontally, the adaxial ones to 4–5 mm from corolla base, the abaxial ones to 3.5–4.5 mm; sparsely hairy. **Host**: *Salvia rosmarinus* (Lamiaceae). **Ecology and distribution**: karstic, slightly sloped, rocky plateaus in the Balearic Islands, Spain and Sardinia (var. *sarda*). **Key diagnostics**: erecto-patent to patent flowers, calyx with long-acuminate triangular teeth, corolla tube uniformly curved; lobes obtuse with a distinctive mucro, some staminal filaments pubescent in the lower ½, anthers hairy; parasitic on rosemary. **Similar species**: considered by some to be a form of *P. rosmarina*; differs subtly in its slenderer stem, 2–4.5 mm diameter at medium height; floral bracts with longer glandular hairs; linear-lanceolate, narrow bracteoles, more acuminate calyx-teeth with denser glandular indumentum, and glandular ciliate corolla margins.

Phelipanche rumseyana
Menorca, Spain (right)

NAME ***Phelipanche schultzii*** (Mutel) Pomel

SYNONYMS ***Orobanche schultzii*** Mutel; ***Phelypaea pulchra*** Pomel; ***Phelypaea schultzii*** (Mutel) Walp.; ***Kopsia schultzii*** (Mutel) Bég.

Habit: rather robust, many-flowered, elongated inflorescence 10–20 cm, cylindrical and gradually narrowing towards the apex; flowers rather lax. **Height**: 10–40(–65) cm. **Stem**: simple (rarely branched), slender to stout, yellowish-brown, sometimes purplish; scales 10–18 mm, ovate or lanceolate, with glandular hairs. **Calyx**: teeth narrowly linear-lanceolate with subulate apex, teeth considerably longer (2–3 ×) than their tube, with abundant glandular hairs; 8–12 mm. Bracteoles linear-lanceolate, shorter than calyx; floral bracts 10–15 mm, lanceolate, with glandular hairs. **Corolla**: sub-erect, some of the basal flowers occasionally pedicellate; tubular to narrowly campanulate, erecto-patent, constricted at the middle, somewhat curved, with glandular hairs; 16–21 mm; violet-blue, lilac, lighter towards the base. Upper lip 2-lobed, lobes usually acute, deflexed; lower lip lobes elliptical, acute or acuminate, with white folds; hairy. **Carpel**: stigma bilobed, whitish; style sub-glabrous. **Stamens**: filaments inserted 4–5 mm above the base of the corolla; ± glabrous; anthers villous, woolly. **Host**: Apiaceae, especially *Ferula* and *Thapsia.* **Ecology and distribution**: xerothermic habitats, rocky slopes and roadsides. Flowering May–July. Mediterranean Basin (Algeria, Morocco, Italy, Portugal, Spain, Canary Islands and Tunisia). **Key diagnostics**: robust plant with long, narrowly linear-lanceolate calyx-teeth, usually 2–3 × longer than the tube. Parasitic on Apiaceae. **Similar species**: other species with long linear-lanceolate and subulate calyx-teeth include *P. heldreichii*, *P. libanotica* and *P. schultzioides* (according to some authors, *P. schultzioides* and *P. gussoneana* are conspecific). *Phelipanche schultzii* is more robust (up to 65 cm) and differs in floral morphology (e.g. calyx with teeth linear-subulate or filiform and 2–3 times longer than tube), and in its host range.

Phelipanche schultzii
Sardinia, Italy (left, page 580, 581)
Portugal (right)

NAME ***Phelipanche sevanensis*** Piwow., Ó.Sánchez & Moreno Mor.

Habit: robust; inflorescence (4–)13–14(–23) cm, cylindrical, shorter than or rarely as long as its stem, usually many-flowered, ± dense. **Height**: (20–)30–33(–37) cm. **Stem**: robust (with 2–5 smaller branches growing from the central part of the shoot), rarely simple, slender; yellowish, usually concolourous with bracts and calyx; in the middle and upper part ± densely glandular-pubescent; scales (10–)11–13(–15) mm, ovate-lanceolate (triangular) to lanceolate, the uppermost densely glandular-pubescent. **Calyx**: with 4 triangular teeth equalling the calyx tube, narrowly lanceolate with subulate to filiform apices, veined, usually exceeding the calyx tube, but shorter than the corolla tube, densely glandular-hairy; (7–)9–11(–13) mm long. Floral bracts (10–)12–14(–17) mm triangular-lanceolate to lanceolate, usually shorter than the calyx. Bracteoles (7–)10–11(–12) mm, shorter than calyx, linear-lanceolate **Corolla**: erect, tubular-infundibuliform, widened at the mouth, dorsal line slightly curved; base slightly inflated, constricted above stamen insertion; (19–)21–23(–25) mm long, 4–5 mm in diameter in the central part; whitish to pale violet, lobes blue to blue-violet with violet veins inside; on the exterior ± densely glandular-pubescent (particularly at the upper lip). Upper lip bilobed, lobes ± erect, with subacute apex, pubescent; lower lip lobes deflexed, the central lobe clearly larger than laterals and elongated, large and broadly ovate with acute apex, sometimes slightly emarginate, ciliate with short glandular hairs and dense white hairs on the folds. **Carpel**: stigma quadrangular to discoid, slightly bilobed, white; style with sparse glandular hairs. **Stamens**: filaments obliquely inserted at 5–6(–7) mm above the corolla base, glabrous or with scarce hairs; anthers distinctly long, hairy. **Host**: *Heracleum trachyloma* (Apiaceae). **Ecology and distribution**: known from one locality, where it grows in rocky, semi-arid mountain steppe on steep slopes at 1,945 m, surrounding Lake Sevan in Armenia. A Caucasian endemic. Flowering end June–July, fruiting July–August. **Key diagnostics**: stem branched, with 2–5 smaller branches growing from the central part of the shoot, or rarely simple, yellowish, usually concolourous with bracts and calyx; floral bracts usually shorter than the calyx, corolla erect, tubular-infundibuliform, whitish to white-violet with darker lobes, blue to blue-violet inside; anthers distinctly long, hairy; peculiar host. **Similar species**: similar to *P. heldreichii*, which is taller, with 2–8 smaller branches growing from the base of the stem or rarely simple, violet to blue, rarely yellowish, usually contrasting with the bracts and calyx; inflorescence longer; corolla tubular (slightly infundibuliform distally) and shorter, ± uniformly violet to blue, with narrower ovate lobes; anthers usually glabrous; calyx-teeth 1–2 × as long as tube; parasitic on *Eryngium*.

Phelipanche sevanensis
Armenia (left, right)

NAME ***Phelipanche zangezuri*** Piwow., Ó.Sánchez & Moreno Mor.

Habit: slender or slightly robust; inflorescence (9–)10–12 cm, cylindrical with an acute apex and shorter than its stem, usually ca. 20-flowered, ± dense. **Height**: (25–)28–30(–32) cm. **Stem**: simple, relatively slender, ± densely glandular-pubescent in the middle and upper portions, with white, very dense and conspicuous glandular hairs; violet or greyish-blue; almost ½ the stem usually submerged; scales (11–)15–19(–21) mm, lanceolate, glandular-pubescent. **Calyx**: with 4 triangular teeth equalling their tube, narrowly lanceolate with a subulate to filiform apex, veined, usually slightly shorter than corolla tube; densely glandular-hairy; tube (6–)7–8 mm long and 6–7 mm wide; teeth (4–)5–6(–7) mm long, (1.0–)1.5–2.0 mm wide at widest point. Floral bracts (16–)17–18 mm, lanceolate, longer than the calyx, with dense white glandular hairs. Bracteoles (9–)10–12 mm, shorter than calyx, linear-lanceolate. **Corolla**: erect and tubular-infundibuliform distally; tube narrow; dorsal line considerably bent in the lower part, almost straight in middle, scarcely widened at tip; (24–)28–30(–31) mm, 5–6 mm in diameter in central part; dark blue or dark violet, lighter (whitish-yellow) towards base of tube; lobes deep blue to blue-violet with violet veins; externally in all its parts very densely glandular-pubescent with white glandular hairs. Upper lip bilobed, ± emarginated and with erect lobes, very densely pubescent; lower lip lobes longer and deflexed, lobes 5–7 mm long, central clearly bigger than the laterals, elongated, acute, slightly emarginated, very densely pubescent, with dense, short white hairs on folds. **Carpel**: stigma bilobed with 2 ovoid, yellowish-white or yellow lobes; style densely glandular-hairy. **Stamens**: filaments inserted 5–6(–7) mm above corolla base, densely non-glandular and glandular-hairy at the base, glabrous or sparsely hairy above; anthers glabrous or sometimes a single line of long hairs along suture. **Host**: *Tragopogon pterocarpus* (Asteraceae). **Ecology and distribution**: slopes and landslides with sparse vegetation, at ca. 2,300 m. Flowering end May–June, fruiting June–July. Known only from southern Armenia. **Key diagnostics**: dense and conspicuous glandular indumentum; inflorescence with acute apex; flowers erect and tubular-infundibuliform distally; corolla long, up to 30 mm, dark blue or dark violet, externally in all its parts very densely glandular-pubescent; lower lip longer and deflexed, lobes long and elongated; floral bracts longer than calyx; calyx-teeth slightly shorter than tube. **Similar species**: other hairy, unbranched species include *Phelipanche pulchella*, *P. hajastanica* or *P. caesia*. *Phelipanche zangezuri* differs from these in the characters stressed above, and in its ecology and host.

Phelipanche zangezuri
Armenia (left, right)

NAME ***Phelypaea boissieri*** (Reut.) Stapf
SYNONYMS ***Anoplanthus biebersteinii* var. *boissieri*** Reut.; ***Diphelypaea boissieri*** (Reut.) Nicolson; ***Diphelypaea coccinea* var. *boissieri*** (Reut.) Uhlich

Habit: clumped with several stems arising from the base with terminal flowers. **Height**: 15(–20) cm. **Stem**: reddish-brown with dense indumentum of minute, brown-tipped glandular hairs, often also glabrous; scales few, darker in colour, ovate and broad-based. **Calyx**: somewhat zygomorphic, weakly 2-lipped, 5-lobed; upper lip formed of 3 erect lobes, all lobes about 1.8–2 cm in length, oblong and tapering to a blunt tip, usually glabrous. **Corolla**: small, cup-shaped tube with 5 strongly overlapping, slightly crenate lobes; 6–8.5 cm wide; exterior of tubes and lobes red or pale, bright orange-yellow; interior orange to scarlet except for two dense black patches at the base of the lowermost lobe into the tube. **Carpel**: stigma 0.3 cm in diameter, pale red; style stout, tapering, pale red. **Stamens**: filaments broad; anthers hairless or sparsely hairy. **Host**: *Centaurea* spp., e.g. *C. macrocephala* and *C. salonitana*, as well as occasionally *Cousinia* (Asteraceae). **Ecology and distribution**: north-west Balkan peninsula and western Asia; Albania, Greece, North Macedonia, Turkey, Iran and Iraq; grassy and rocky mountain slopes (even in elevation 2,500 m), steppe. Flowering May–June (July). **Key diagnostics**: markedly overlapping petals. **Similar species**: *Phelypaea coccinea* lacks the overlapping petals and hairless anthers, although molecular and morphological characteristics do not differentiate these taxa entirely; further research into their variability is required. *Phelypaea tournefortii* has long, multicellular, white hairs along the stems and calyx, and different hosts.

Infraspecific taxa

Forma *lutea* differs in colour (corolla, calyx and scales are yellow to orange with black folds in the corolla, with only the stem being brownish) and, usually, in having a longer stem. So far known only from south-east Turkey.

Phelypaea boissieri
Greece (left, right)
photos: Attila Mátis
Phelypaea boissieri f. *lutea*
Turkey (page 588, 589)
photos: Ümit Subaşı

NAME ***Phelypaea coccinea*** (M.Bieb.) Poir.

SYNONYMS ***Anoplanthus coccineus*** (M.Bieb.) Walp.; ***Anoplon coccineum*** (M.Bieb.) Riedl & Schiman-Czeika; ***Diphelypaea coccinea*** (M.Bieb.) Nicolson; ***Orobanche coccinea*** M.Bieb.; ***Orobanche phelypaea*** M.Bieb.; ***Phelypaea biebersteinii*** Fisch. ex Walp.

Habit: simple or clumped with several stems arising from the base with terminal flowers. **Height**: up to 30 cm. **Stem**: reddish-brown, glabrous or almost glabrous; scales 2–5 evenly distributed along the stems, ovate, oblong and semi-clasping the stems. **Calyx**: broadly bell-shaped, unevenly 5 lobed; lobes oblong, glabrous. **Corolla**: tube curved, with 5 round-oblong lobes, rarely overlapping; tube longer than wide, 1.5–2.5 cm long; exterior scarlet or yellowish; interior of tubes and lobes scarlet except for two dense black patches at the base of the lowermost lobe into the tube (occasionally entirely yellow with black folds). **Carpel**: stigma 0.3 cm in diameter, pale red, rarely yellow; style stout, tapering, pale red. **Stamens**: anthers hairless and minute, 0.3 cm long. **Host**: *Psephellus* spp. including *P. amblyolepis*, *P. dealbatus*, *P. carthalinicus* and *Klasea* spp. including *K. radiata* and *K. coriacea*, rarely *Centaurea pseudoscabiosa* (Asteraceae). **Ecology and distribution**: mainly Caucasus (Russia, Georgia, Armenia, Azerbaijan, Turkey) and Crimea; records in the Balkan peninsula probably correspond with *P. boissieri*; grassy or rocky mountain slopes, steppe, semi-desert hills, edges of forest, at usually 800 to 2,300 (2,900) m. Flowering May–June (July). **Key diagnostics**: glabrous plant with scarlet, dark-throated flowers with usually obovate and scarcely overlapping lobes, scales along the stem and hairless anthers. **Similar species**: *Phelypaea boissieri* has markedly overlapping corolla lobes; *P. tournefortii* has a cup-shaped tube corolla, basal scales and long multicellular white hairs on the calyx and along the stems, as well as a different host range.

Phelypaea coccinea Armenia (above), Georgia (page 592, 593), illustrated (right)

NAME ***Phelypaea tournefortii*** Desf.
SYNONYMS ***Anoplon tournefortii*** (Desf.) G.Don; ***Diphelypaea tournefortii*** (Desf.) Nicolson

Habit: clumped with several stems arising from the base with terminal flowers. **Height**: 20–30 cm. **Stem**: reddish-brown, glabrous or almost glabrous. Scales short and broadly ovate, located towards the stem base. **Calyx**: clearly bilabiate, 1.5–2.5 cm long; tube is 0.7–1.0 cm long; lobes oblong, hairy. **Corolla**: tube short and upturned, cup-shaped with 5 widely spreading, strongly overlapped, slightly crenate lobes, wider than long; interior and exterior of tubes and lobes crimson except for two dense black patches at the base of the lowermost lobe into the tube. Upper lip lobes, about equal to the lower lip or a little shorter, 1–2.5 cm in diameter. **Carpel**: stigma 0.3 cm in diameter, pale red, rarely yellow; style stout, tapering, pale red. **Stamens**: filaments white-hairy; anthers hairy. **Host**: *Tanacetum*, mainly *T. polycephalum*, *T. chilophyllum* and *T. canescens* (Asteraceae). **Ecology and distribution**: mainly Transcaucasia (Armenia and Azerbaijan) and Turkey; rocky slopes, steppe, usually at (1,200–)1,800–2,100(–2,500) m. Flowering May–June. **Key diagnostics**: white indumentum on the stem and calyx and scales concentrated towards the stem base. **Similar species**: *Phelypaea coccinea* has a tube-like corolla, hairless anthers, calyx and stem, lobes not overlapping, scales located along the stem; *P. boissieri* has markedly overlapping petals as well as different hosts; neither species possesses the white, long, multicellular-hairy indumentum of *P. tournefortii.*

Phelypaea tournefortii Armenia (above, page 596, 597), illustrated (right)

NAME *Xylanche himalaica* (Hook.f. & Thomson) Beck
SYNONYM *Boschniakia himalaica* Hook.f. & Thomson

Habit: tall, robust, rootstock globose or sub-globose, stems usually 1, inflorescences racemose, many-flowered, erect, solitary or few stems. **Height**: 15–40 cm. **Stem**: non glandular-hairy with numerous, rigid, lanceolate scales, overlapping at the stem base; brown. **Calyx**: short, cup-shaped, 4–5 × 5–8 mm, 5-lobed, the lobes unequal, linear-lanceolate or narrowly triangular, 2–5 mm, usually partly or all deciduous after anthesis and calyx margin entire. **Corolla**: tube slightly enlarged, 1.5–2.5 cm; upper lip emarginate or entire, 7–9 mm; lower lip 2–3 mm, 3-lobed; lobes pubescent at margin; yellow-brown or pale purple. **Carpel**: stigma usually 3-lobed; style glabrous; seeds irregularly globose. **Stamens**: filaments 1.2–1.4 cm, basally pubescent, glabrescent above; anthers ovoid-oblong, glabrous. **Host**: *Rhododendron* spp. (Ericaceae). **Ecology and distribution**: central Himalaya to western and central China, Taiwan; in forests and above treeline in the rhododendron bushes, usually at elevation 2,500–4,500(–4,800) m. Flowering April–June (July). **Key diagnostics**: sombre-coloured spikes with hooded, open-mouthed flowers are unmistakable. **Similar species**: *Boschniakia rossica*, its sister species, is somewhat similar but has long cylindric rootstock, usually with more stems, spicate inflorescences, corolla smaller (1–1.2 cm) with different morphology and colour, shorter pedicel, ellipsoid-globose seeds, and a different distribution and host range.

Xylanche himalaica Taiwan (above, right) photos: Mu-Lin Chang

GLOSSARY

Abaxial, surface facing away from the axis (underside).

Actinomorphic, floral parts with a regular radial symmetry.

Acuminate, tapering to a long point.

Adaxial, surface facing towards the axis (upperside).

Allopatric, a population or species that is physically isolated from other similar groups by an extrinsic barrier.

Angiosperm, flowering plant.

Annual, completing the life cycle in one year.

Anther, fertile, pollen-producing part of the stamen, typically on a terminal stalk (**filament**).

Antrorse, directed forward.

Apex, uppermost part of a structure.

Apiculate, having a short, sharply pointed tip.

Arachnoid-woolly, with a web of hairs.

Attenuate, tapering gradually to a point.

Autotrophic, an organism that produces its own food.

Bidentate, divided into two teeth.

Biennial, completing its life cycle in two years.

Bifid, divided into two parts, typically deeply at the apex.

Bilabiate, divided into two lips.

Bilamellate, with two elevated ridges.

Bilaminar, consisting of two thin structures.

Bilobed, divided into two lobes.

Bract, small, leaf-like structure, often subtending (beneath) the flower (floral bract).

Bracteole, small secondary bract.

Calceiform, slipper-like.

Calyx, all the sepals of the flower.

Campanulate, bell-shaped.

Canescent, hoary – covered with short white or greyish hairs.

Capitate, rounded or head shaped.

Capsule, a dry fruit opening when ripe.

Carinate, keeled or ridged.

Carpel, female part in a flower, made of the ovary, the stigma, and the style.

Cernuous, drooping or nodding.

Chasmogamous, flowers open for cross-fertilisation, cf. **cleistogamous**.

Ciliate, with fine hairs on the margin.

Clavate, club shaped, slender and distally thickened.

Cleistogamous, self-fertilised within a permanently closed flower.

Comose, with a tuft of hairs.

Concolourous, of the same colour throughout.

Connate, partially joined together.

Corolla, all the petals of a flower (which may form a tube).

Corymbose, having the form of a corymb – an inflorescence, flat topped as the outer pedicels are longer.

Crateriform, saucer shaped

Crenate, with round-toothed or scalloped edges.

Crenulate, having a finely notched or scalloped edge.

Cucullate, hooded, shaped like a cowl.

Cymose, having the form of a cyme – an inflorescence in which each floral axis terminates in a single flower.

Decumbent, sprawling but with the apex turning upwards.

Deflexed, bent downwards.

Deltate, triangular.

Dentate, toothed.

Cistanche deserticola
Inner Mongolia (left)

Denticulate, finely toothed.

Didynamous, having four stamens of unequal length.

Discoid, shaped like a disk.

Distal, at the end furthest from the point of attachment, cf. **proximal**.

Divided, not entire; divided into teeth, lobes or leaflets.

Eglandular, lacking glands.

Ellipsoid, elliptic in outline.

Emarginate, notched at the apex.

Endoparasite, parasite whose vegetative structures are all internal in the host.

Entire, whole; without distinct lobes, teeth or divisions.

Epipetalous, borne on the petals or corolla.

Erecto-patent, between erect and spreading.

Exserted, protruding (not included, such as anthers from a corolla tube).

Facultative, not obliged to follow a single option: a parasite capable of completing its life cycle in the absence of a host, cf. **obligate**.

Ferruginous, rust-coloured.

Filament, the stalk of the stamen.

Galeate, hooded.

Gamosepalous, with united or fused sepals.

Geniculate, bent abruptly.

Genus (pl. **genera**)**,** monophyletic group of related species, the taxonomic group between the lower rank of species and the higher rank of family; the generic name is the first part of the scientific binomial.

Geoflorous, flowering at ground-level.

Gibbous, fattened or bulging.

Glabrescent, almost glabrous.

Glabrous, not hairy, smooth, lacking hairs.

Gland, organ of secretion, often in sticky plants; adj. **glandular/glanduliferous** (often referring to hairs).

Globose, spherical.

Halophyte, plant adapted to grow in saline conditions; adj. **halophytic**.

Haustorium, specialised root-like structure penetrating the host and drawing water and nutrients from it.

Head, group of flowers crowded together at the end of a stalk.

Hemiparasite, parasitic plant that extracts some of its nutrition from another plant (the host) but which produces chlorophyll.

Herb, plant without woody parts; a soft and leafy annual, biennial or perennial in which aerial parts naturally die to ground-level at the end of the growing season; adj. **herbaceous**.

Heterotrophic, organism that cannot manufacture its own food by carbon fixation and therefore derives its intake of nutrition from other sources of organic carbon.

Holoparasite, parasitic plant that extracts all its nutrition from another plant (the host) and which lacks chlorophyll and a true root system.

Horizontal gene transfer (HGT), non-sexual DNA transmission between distantly related organisms that avoids the cross-species barriers.

Host specificity, the diversity of host species that can be infected by a parasite. Highly **host-specific** parasites occur on a single host species; **generalist** parasites infect a range of species.

Host, the plant infected by the parasite.

Hybrid, offspring of a cross between two or more different species, races or varieties.

Hybridisation, the formation of hybrid offspring.

Imbricate, overlapping like tiles on a roof.

Indumentum, collective name for the hairs and trichomes arising from the epidermal surface.

Inflected, bent or folded downward.

Infundibuliform, funnel shaped.

Lacerate, tattered.

Lanate, thickly woolly-hairy.

Lanceolate, lance-shaped; tapering to a point at the apex and base.

Life history, the lifespan and various phases from germination to maturity.

Lip, region of the corolla sharply differentiated from the rest (upper and lower).

Lobe, substantial division of the calyx or corolla.

Locule, a cavity or compartment within an organ, usually referring to the ovary.

Monocarpic, dying after setting seed.

Monothecous, anther with a single **theca** or pollen-releasing compartment.

Morphology, the appearance, form or structure.

Mucro, a short, sharp point; adj. **mucronate**.

Mycoheterotrophy, life history by which a non-photosynthetic plant obtains nutrition from a fungal symbiont (or sometimes from another plant via a shared fungal symbiont) living in its root system.

Mycorrhiza, mutually beneficial relationship between the root of a plant and a fungus that colonises the plant root.

Native, naturally occurring in the area (not **introduced** or **naturalised**).

Obligate, parasite unable to complete its life cycle in the absence of a host, cf. **facultative**.

Ochroleucous, yellowish-white.

Ovate, oval/egg-shaped, with a tapering point and widest near the point of attachment.

Palatal, of the palate, a projecting part of the lip closing the mouth of the corolla.

Panicle, a compound, or much-branched inflorescence.

Papillate, pimple covered.

Papillose-verrucose, pimply to warty.

Parasite, an organism that derives nutrition from another organism (its **host**).

Patent, spreading.

Pedicel, individual flower stalk.

Peltate, shield shaped with a stalk arising from its plane surface.

Perennial, living for more than two years, generally flowering every year. Often woody at the base.

Photoautotrophic, organism producing its own nutrition through photosynthesis.

Polyploid, an organism having more than two sets of homologous chromosomes.

Porrect, directed outwards and forwards and somewhat upswept.

Proximal, at the end near the point of attachment, cf. **distal**.

Puberulous/puberulent, covered with minute, short, soft hairs.

Pubescent, downy with short, soft hairs.

Raceme, an unbranched inflorescence in which the flowers are borne on **pedicels.**

Racemose, having the form of a **raceme**.

Radicle, embryonic root structure that initiates host contact.

Reflexed, bent back on itself.

Reniform, kidney-shaped.

S.l., abbreviation of *sensu lato* (in the broad sense), usually in reference to a species.

S.str., abbreviation of *sensu stricto* (in the narrow sense), usually in reference to a species.

Sagittate, shaped like an arrowhead.

Scale, reduced leaves in holoparasites found at the base of the stem.

Serrate, coarsely toothed, saw-like.

Sessile, not stalked.

Sinuate, having a wavy (sinuous) margin.

Spathe, a sheathing bract enclosing the floral parts.

Spathulate, spoon shaped.

Species, the fundamental unit of classification; a species name comprises two words: the **genus** name (e.g. *Orobanche*) and a specific epithet (e.g. *minor*).

Spicate/spiciform, in the form of a **spike**.

Spike, a **racemose** inflorescence in which the flowers are **sessile**.

Squamiform, resembling a scale.

Stamen, male reproductive organ of the flower consisting of a filament and anther (microsporangium).

Stigma, the part of the carpel or pistil that receives pollen and upon which the pollen germinates.

Striated, narrowly grooved or striped.

Subspecies, the taxonomic rank below the species level.

Subulate, narrow, broadest at the base, and tapered to an elongated tip.

Taxon (pl. **taxa**)**,** taxonomic unit of any rank, for example species, genus, subspecies or variety.

Teeth, divisions of a leaf, calyx or corolla; adj. **toothed**.

Theca (pl. **Thecae**)**,** pollen-releasing compartments of the anther, typically paired.

Thyrsoid, in the form of a thyrse, a contracted ovate **panicle** with **cymose** branches.

Trichome, small hair or other outgrowth, typically unicellular and glandular.

Tubercle, fleshy, root-like starch-storage structure.

Urceolate, urn shaped.

Variety, subgroup of a species that is morphologically and/or genetically distinct (but not distinct enough to warrant subspecies status).

Villose/villous, covered with long, soft, straight hairs.

Woolly, clothed with soft, shaggy hairs.

Xerothermic, dry, or seasonally prone to drought.

Zygomorphic, bilaterally symmetrical floral arrangement.

SELECTED REFERENCES AND FURTHER READING

Aldughayman, M., Thorogood, C.J., Al-Mayah, A., Abdulridha, A. & Hawkins, J.A. 2024. An account of the genus *Cistanche* (Orobanchaceae) in Iraq and taxonomic considerations in the Middle East. *PhytoKeys* 238: 281–294.

Atkinson, M. 1996. The distribution and naturalisation of *Lathraea clandestina* L. (Orobanchaceae) in the British Isles. *Watsonia* 21: 119–128.

Beck, G. 1890. Monographie der Gattung *Orobanche*. *Bibliotheca Botanica* 19: 1–275.

Beck, G. 1930. Orobanchaceae L. In: Engler, A. (Ed.), *Das Pflanzenreich IV. 261 (Heft 96)*. Wilhelm Engelmann, Leipzig, pp. 1–348.

Ben-Natan, D., Thorogood, C.J. 2025. A revision of the genus *Cistanche* (Orobanchaceae) in Israel, and considerations for its taxonomic circumscription in the Middle East. *PhytoKeys* 260: 213-236.

Bennett, J. & Mathews, S. 2006. Phylogeny of the parasitic plant family Orobanchaceae inferred from phytochrome A. *American Journal of Botany* 93: 1039–1051.

Boissier, P.É. 1879. *Flora orientalis [...] Volumen quartum. Fasciculus secundus*. Genevae et Basileae.

Britten, J. 1881. *The Names of Herbes by William Turner, A.D. 1548*. N. Trübner & Co., London.

Carlón, L., Gómez Casares, G., Laínz, M., Moreno Moral, G., Sánchez Pedraja, Ó. & Schneeweiss, G.M. 2005. Más, a propósito de algunas *Orobanche* L. y *Phelipanche* Pomel (Orobanchaceae) del oeste del Paleártico. *Documentos Jardín Botánico Atlántico* (Gijón) 3: 1–71.

Carlón, L., Gómez Casares, G., Lainz, M., Moreno Moral, G., Sánchez Pedraja, Ó. & Schneeweiss, G.M. 2008. Más, a propósito de algunas *Phelipanche* Pomel, *Boulardia* F.W. Schultz y *Orobanche* L. (Orobanchaceae) del oeste del Paleártico. *Jardín Botánico Atlántico* 6: 1–128.

Carlón, L., Laínz, M., Moreno Moral, G. & Sánchez Pedraja, Ó. 2011. A new species (*Orobanche loscosii*), a prioritary name for *O. icterica* (*O. leptantha*) and a new member of the Spanish flora (*O. elatior*). *Flora Montiberica* 48: 89–101.

Carlón, L., Laínz, M., Moreno Moral, G. & Sánchez Pedraja, Ó. 2015. The overlooked *Orobanche balsensis* (J.A. Guim.), comb. nov., and some remarks on *O. subbaetica* Triano & A. Pujadas. *Flora Montiberica* 60: 38–53.

Castillo-Sánchez, I.L. & Figueroa-Castro, D.M. 2019. Intra-inflorescence variation in reproductive traits of *Conopholis alpina* (Orobanchaceae): effect of flower maturation pattern and resource competition. *Plant Ecology* 220: 721–729.

Chater, A.O. & Webb, D.A. 1972. *Orobanche* L. In: Tutin, T.G. et al., *Flora Europaea* vol. 3, pp. 286–293. Cambridge University Press, Cambridge, UK.

Collins, L.T., Velazco, C. & Schneider, A.C. 2023. Two new species of *Aphyllon* from northeastern Mexico. *Systematic Botany* 48 (4): 587–594.

Cullen, J. 2010. *Phelypaea boissieri*: Orobanchaceae. *Curtis's Botanical Magazine* 26 (4): 379–388.

Domina, G. 2018. Host-driven morphological variability in *Orobanche crenata* (Orobanchaceae). *Turkish Journal of Botany* 42: 502–509.

Domina, G. & Mazzola, P. 2011. Notes on the genus *Orobanche* in Italy: 3. Taxa. *Plant Biosystems* 145: 342–346.

Domina, G. & Mazzola, P. 2011. The genus *Orobanche* in Sicily. *Flora Mediterranea* 21: 205–242.

Domina, G. & Soldano, A. 2015. *Orobanche apuana* (Orobanchaceae): a new species endemic to Italy. *Phytotaxa* 207: 163–171.

Orobanche javakhetica
Georgia (left)

Domina, G. & Stepanek, J. 2009. Typification of the name *Orobanche canescens* C. Presl (Orobanchaceae) with taxonomic notes. *Candollea* 64: 31–37.

Domina, G., Uhlich, H. & Barone, G. 2022. *Orobanche australis* Moris ex Bertol. the correct name for *O. thapsoides* Lojac (Orobanchaceae). *Phytotaxa* 531: 91–96.

Foley, M.J.Y. 2001a. Orobanchaceae in the "Flora Iberica" area: new taxa, excluded taxa, and typification. *Anales Jardín Botánico Madrid* 58 (2): 223–233.

Foley, M.J.Y. 2001b. *Orobanche* L. In: Paiva, J., Sales, F., Hedge, I.C., Aedo, C., Aldasoro, J.J., Castroviejo, S., Herrero, A. & Velayos, M. (Eds.), *Flora Iberica* 14: 32–72. Madrid.

Freeman, C.C., Rabeler, R.K. & Elisens, W.J. 2019. Orobanchaceae. In: *Flora of North America North of Mexico*. Volume 17: *Magnoliophyta: Tetrachondraceae to Orobanchaceae*. Oxford University Press, New York, pp. 465–768.

Fu, Z., Fan, X., Wang, X. & Gao, X. 2018. *Cistanches Herba*: an overview of its chemistry, pharmacology, and pharmacokinetics property. *Journal of Ethnopharmacology* 219: 233–247.

Fu, W., Liu, X., Zhang, N., Song, Z., Zhang, W., Yang, J. & Wang, Y. 2017. Testing the hypothesis of multiple origins of holoparasitism in Orobanchaceae: phylogenetic evidence from the last two unplaced holoparasitic genera, *Gleadovia* and *Phacellanthus*. *Frontiers in Plant Science* 8: 1380.

Gerard, J. 1597. *The Herball or Generall Historie of Plantes*. John Norton, London.

Gilli, A. 1966. Bestimmungsschlüssel der Mitteleuropäischen Varietäten und Formen von *Orobanche*. *Verhandlungen der Zoologisch-Botanischen Gesellschaft in Wien* 105–106: 171–181.

Gilli, A. 1968. *Necranthus*: a new genus of Orobanchaceae from Turkey. *Notes from the Royal Botanic Garden, Edinburgh* 28 (3): 297–298.

Gilli, A. 1974. Orobanchaceae. In: Hegi, G. (Ed.), *Illustrierte Flora von Mitteleuropa*, ed. 2, 6. pp. 470–505. München.

Gilli, A. 1982. Orobanchaceae (90). In: Davis, P.H. (Ed.). *Flora of Turkey* 7. Edinburgh University Press, pp. 1–22.

Gracie, C. 2012. *Spring Wildflowers of the Northeast: A Natural History*. Princeton University Press, pp. 173–177.

Hatt, S., Tsiftsis, S. & Thorogood, C.J. 2024. A taxonomic monograph of the Eurasian holoparasitic plant genus *Lathraea* L. (Orobanchaceae). *Phytotaxa* 672 (1): 1–29.

Haynes, R.R. 1971. A monograph of the genus *Conopholis* (Orobanchaceae). *SIDA, Contributions to Botany* 4 (3): 246–264.

Heckard, L.R. 1973. A taxonomic re-interpretation of the *Orobanche californica* complex. *Madrono* 22: 41–70.

Heide-Jørgensen, H.S. 2008. *Parasitic Flowering Plants*. Brill, The Netherlands.

Joel, D.M., Gressel, J. & Musselman, L.J. (Eds.) 2013. *Parasitic Orobanchaceae: Parasitic Mechanisms and Control Strategies*. Springer, Heidelberg, Germany.

Johnson, S.D. & Raguso, R.A. 2016. The long-tongued hawkmoth pollinator niche for native and invasive plants in Africa. *Annals of Botany* 117 (1): 25–36.

Kreutz, C.A.J. 1995. *Orobanche. The European Broomrape Species. Central and Northern Europe*. Natuurhistorisch Genootscha, Limburg.

Kuijt, J. 1969. *The Biology of Parasitic Flowering Plants*. University of California Press, Berkeley, CA.

Kumar, A., Page, N.V., Adhikari, B.S., Nair, M.V. & Rawat, G.S. 2021. On the rediscovery of a rare root parasite *Gleadovia ruborum* Prain & Gamble (Orobanchaceae) from Uttarakhand, western Himalaya, India. *Journal of Threatened Taxa* 13 (8): 19185–19188.

Lachowicz-Wiśniewska, S., Piwowarczyk, R., Ochmian, I., Kapusta, I., Bernatek, M. & Piątek, J. 2024. Correlated trophic and bioactive activities in the parasite–host relationship – *Phelipanche purpurea* vs. *Achillea arabica* case study. *Industrial Crops and Products* 204, Part B, 15: 117379.

Li, X., Feng, T., Randle, C. & Schneeweiss, G.M. 2019. Phylogenetic relationships in Orobanchaceae inferred from low-copy nuclear genes: consolidation of major clades and identification of a novel position of the non-photosynthetic *Orobanche* clade sister to all other parasitic Orobanchaceae. *Frontiers in Plant Science* 10: 902.

Li, X., Jang, T.S., Temsch, E.M., Kato, H., Takayama, K. & Schneeweiss, G.M. 2017. Molecular and karyological data confirm that the enigmatic genus *Platypholis* from Bonin-Islands (SE Japan) is phylogenetically nested within *Orobanche* (Orobanchaceae). *Journal of Plant Research* 130 (2): 273–280.

Li, Z., Lin, H., Gu, L., Gao, J., & Tzeng, C.-M. 2016. *Herba Cistanche* (*Rou Cong-Rong*): one of the best pharmaceutical gifts of traditional Chinese medicine. *Frontiers in Pharmacology* 7: 41.

Linnaeus, C. 1753. *Species Plantarum*. Stockholm.

Lunau, K., De Camargo, M.G.G. & Brito, V.L.G. 2024. Pollen, anther, stamen, and androecium mimicry. *Plant Biology* 26 (3): 349–368.

McNeal, J.R., Bennett, J.R., Wolfe, A.D. & Mathews, S. 2013. Phylogeny and origins of holoparasitism in Orobanchaceae. *American Journal of Botany* 100: 971–983.

Moreno Moral, G., Sánchez Pedraja, Ó. & Piwowarczyk, R. 2018. Contributions to the knowledge of *Cistanche* (Orobanchaceae) in the Western Palearctic. *Phyton* (Horn, Austria) 57 (1–2): 19–36.

Musselman, L.J. 1980. The biology of *Striga*, *Orobanche*, and other root-parasitic weeds. *Annual Review of Phytopathology* 18: 463–489.

Nickrent, D.L. 2020. Parasitic angiosperms: how often and how many? *Taxon* 69 (1): 5–27.

Niedzwetzki-Taubert, T. 2022. *Hyobanche hanekomii* (Orobanchaceae) is pollinated by non-flying mammals and birds. *bioRxiv* 2022.12.13.519978.

Nishimura, A. & Takayama, K. 2022. First record of potential bird pollination in the holoparasitic genus *Orobanche* L. *Plant Species Biology* 38 (1): 6–17.

Novopokrovsky, I.V. & Tzvelev, N.N. 1958. Orobanchaceae. In: Shishkin, B.K. (Ed.), *Flora Unionis Republicarum Socialisticarum Sovieticarum (Flora URSS)*. Institutum Botanicum nomine V.L. Komarovii Academiae Scientiarum URSS, Mosqua and Leningrad, vol. 23, pp. 685–687.

Olmstead, R.G., DePamphilis, C.W., Wolfe, A.D., Young, N.D., Elisons, W.J. & Reeves, P.A. 2001. Disintegration of the Scrophulariaceae. *American Journal of Botany* 88 (2): 348–361.

On-Nom, N., Thangsiri, S., Inthachat, W., Temviriyanukul, P., Sahasakul, Y., Aursalung, A., Chupeerach, C. & Suttisansanee, U. 2024. Optimized conditions for the extraction of phenolic compounds from *Aeginetia indica* L. and its potential biological applications. *Molecules* 29 (5): 1050.

Park, J.M., Manen, J.F., Colwell, A.E. & Schneeweiss, G.M. 2008. A plastid gene phylogeny of the non-photosynthetic parasitic *Orobanche* (Orobanchaceae) and related genera. *Journal of Plant Research* 121: 365–376.

Pignone, D. & Hammer, K. 2016. Parasitic angiosperms as cultivated plants? *Genetic Resources and Crop Evolution* 63: 1273–1284.

Piwowarczyk, R. 2015a. Seed micromorphology of central European *Orobanche* and *Phelipanche* (Orobanchaceae) in relation to preferred hosts and systematic implications. *Australian Systematic Botany* 28: 124–136.

Piwowarczyk, R. 2015b. Seed morphology of *Boschniakia sensu lato* (Orobanchaceae) and its taxonomical implications. *Phytotaxa* 231 (2): 156–164.

Piwowarczyk, R. & Kasińska, J. 2017. Petal epidermal micromorphology in holoparasitic Orobanchaceae and its significance for systematics and pollination ecology. *Australian Systematic Botany* 30: 48–63.

Piwowarczyk, R. & Kolanowska, M. 2023a. Effect of global warming on the potential distribution of a holoparasitic plant (*Phelypaea tournefortii*): both climate and host distribution matter. *Scientific Reports* 13 (1): 10741.

Piwowarczyk, R. & Kolanowska, M. 2023b. Predicting the effect of global warming on the distribution of a polyphagous tree parasite, *Orobanche laxissima*, based on climatic and ecological data. *Global Ecology and Conservation* 44: e02486.

Piwowarczyk, R., Denysenko-Bennett, M., Góralski, G., Kwolek, D., Sánchez Pedraja, Ó., Mizia, P., Cygan, M. & Joachimiak, A.J. 2018. Phylogenetic relationships within *Orobanche* and *Phelipanche* (Orobanchaceae) from Central Europe, focused on problematic aggregates, taxonomy, and host ranges. *Acta Biologica Cracoviensia. Series Botanica* 60 (1): 45–64.

Piwowarczyk, R., Ruraż, K., Krasylenko, Y., Kasińska, J. & Sánchez Pedraja, Ó. 2020a. Seed micromorphology of representatives of holoparasitic Orobanchaceae genera from the Caucasus region and its taxonomic significance. *Phytotaxa* 432 (3): 223–251.

Piwowarczyk, R., Ochmian, I., Lachowicz, S., Kapusta, I., Sotek, Z. & Błaszak, M. 2020b. Phytochemical parasite–host relations and interactions: a *Cistanche armena* case study. *Science of the Total Environment* 716: 137071.

Piwowarczyk, R., Ochmian, I., Lachowicz, S., Kapusta, I. & Sotek, Z. 2020c. Phytochemical and bioactive properties of *Phelypaea tournefortii* – effect of parasitic lifestyle and environmental factors. *Acta Universitatis Cibiniensis. Ser E: Food Technology* 24 (1): 113–128.

Piwowarczyk, R., Sánchez Pedraja, Ó., Moreno Moral, G., Fayvush, G., Zakaryan, N., Kartashyan, N. & Aleksanyan, A. 2019. Holoparasitic Orobanchaceae (*Cistanche*, *Diphelypaea*, *Orobanche*, *Phelipanche*) in Armenia: distribution, habitats, host range and taxonomic problems. *Phytotaxa* 386 (1): 1–106.

Piwowarczyk, R., Sánchez Pedraja, Ó., Khutsishvili, M. & Kharazishvili, D. 2023. Holoparasitic Orobanchaceae in Georgia (Caucasus): taxonomic revision, diversity, distribution, habitats and host range. *Phytotaxa* 604: 1–103.

Piwowarczyk, R., Schneider, A.C., Góralski, G., Kwolek, D., Denysenko-Bennett, M., Burda, A., Ruraż, K., Joachimiak, A.J. & Sánchez Pedraja Ó. 2021. Phylogeny and historical biogeography analysis support Caucasian and Mediterranean centres of origin of key holoparasitic *Orobancheae* (Orobanchaceae) lineages. *PhytoKeys* 174: 165–194.

POWO, 2025. Plants of the World Online. Facilitated by the Royal Botanic Gardens, Kew. https://powo.science.kew.org/ (Retrieved January 2025).

Pujadas-Salvà, A.J. 2009. *Orobanche* L. In: Blanca, G., Cabezudo, B., Cueto, M., Fernández López, C. & Morales Torres, C. (Eds.), *Flora Vascular de Andalucía Oriental 3. Consejería de Medio Ambiente, Junta de Andalucía*. Sevilla, pp. 412–421.

Pujadas Salvá, A. & Lora González, Á. 1997. The genus *Orobanche* L. (Orobanchaceae) in the province of Almería (SE of the Iberian Peninsula). *Lagascalia* 19 (1–2): 873–880.

Pujadas Salvà, A. & Velasco, L. 2000. Comparative studies on *Orobanche cernua* L. and *O. cumana* Wallr. (Orobanchaceae) in the Iberian Peninsula. *Botanical Journal of the Linnean Society* 134: 513–527.

Pusch, J. & Günther, K.F. 2009. Orobanchaceae (Sommerwurzgewächse). In: Hegi, G. (Ed.), *Illustrierte Flora von Mitteleuropa 6/1A, Lieferung 1*. Weissdorn-Verlag, Jena, 99 pp.

Randle, C.P. 2006. Revision of *Harveya* (Orobanchaceae) of southern Africa. *Systematic Botany Monographs* 80: 1–74.

Renna, M., Serio, F. & Santamaria, P. 2015. Crenate broomrape (*Orobanche crenata* Forskal): prospects as a food product for human nutrition. *Genetic Resources and Crop Evolution* 62: 795–802.

Rumsey, F.J. & Jury, S.L. 1991. An account of *Orobanche* in Britain and Ireland. *Watsonia* 18: 257–295.

Rumsey, F.J. & Thorogood, C.J. 2023. The shifting ecology and distribution of one of Britain's rarest plants: *Orobanche picridis* (Orobanchaceae). *British and Irish Botany* 5 (3): 303–319.

Ruraż, K., Przemieniecki, S.W. & Piwowarczyk, R. 2023. Interspecies and temporal dynamics of bacterial and fungal microbiomes of pistil stigmas in flowers in holoparasitic plants of the *Orobanche* series *Alsaticae* (Orobanchaceae). *Scientific Reports* 13: 6749.

Sánchez Pedraja, Ó., Moreno Moral, G., Carlón, L., Piwowarczyk, R., Laínz, M. & Schneeweiss, G.M. 2016. Index of Orobanchaceae. http://www.farmalierganes.com/Otrospdf/publica/Orobanchaceae%20Index.htm. Liérganes, Cantabria, Spain. ISSN: 2386-9666 (Retrieved January 2025).

Schiman-Czeika, H. 1964. Orobanchaceae. In: Rechinger, K.H. (Ed.), *Flora Iranica (Flora des Iranischen Hochlandes und der umrahmenden Gebirge: Persien, Afghanistan, Teile von West-Pakistan, Nord-Iraq, Azerbaidjan, Turkmenistan)*. Graz-Austria. No. 5: Orobanchaceae, 25 pp., 8 tab.

Schneeweiss, G.M., Colwell, A., Park, J.M., Jang, C.G. & Stuessy, T.F. 2004a. Phylogeny of holoparasitic *Orobanche* (Orobanchaceae) inferred from nuclear ITS sequences. *Molecular Phylogenetics and Evolution* 30 (2): 465–478.

Schneeweiss, G.M., Palomeque, T., Colwell, A.E. & Weiss-Schneeweiss, H. 2004b. Chromosome numbers and karyotype evolution in holoparasitic *Orobanche* (Orobanchaceae) and related genera. *American Journal of Botany* 91: 439–448.

Schneider, A.C. 2016. Resurrection of the genus *Aphyllon* for New World broomrapes (*Orobanche* s.l., Orobanchaceae). *PhytoKeys* 75: 107–118.

Schneider, A.C. & Moore, A.J. 2017. Parallel Pleistocene amphitropical disjunctions of a parasitic plant and its host. *American Journal of Botany* 104 (11): 1745–1755.

Shi, R., Zhang, C., Gong, X., Yang, M., Ji, M., Jiang, L., Leonti, M., Yao, R. & Li, M. 2020. The genus *Orobanche* as food and medicine: an ethnopharmacological review. *Journal of Ethnopharmacology* 263: 113154.

Teryokhin, E.S. 1997. *Weed Broomrapes, Systematics, Ontogenesis, Biology, Evolution*. Aufstieg-Verlag, Landshut.

Teryokhin, E.S., Schibakina, G.V., Serafimovitsch, N.B. & Kravtzova, T.I. 1993. *Opredelitel' Zarazichovych Flory SSSR*. Nauka, Sankt Petersburg.

Těšitel, J., Li, A.R., Knotková, K., McLellan, R., Bandaranayake, P.C.G. & Watson, D.M. 2021. The bright side of parasitic plants: what are they good for? *Plant Physiology* 185 (4): 1309–1324.

Thieret, J.W. 1971. The genera of Orobanchaceae in the southeastern United States. *Journal of the Arnold Arboretum* 52: 404–434.

Thorogood, C.J. & Rumsey, F.J. 2020. An account of common broomrape *Orobanche minor* (Orobanchaceae) in the British Isles. *British & Irish Botany* 2 (3): 223–239.

Thorogood, C.J. & Rumsey, F.J. 2021. *Broomrapes of Britain and Ireland*. Botanical Society of Britain and Ireland. UK.

Thorogood, C.J., Leon, C.J., Lei, D., Aldughayman, M., Huang, L.-F. & Hawkins, J.A. 2021. Desert hyacinths: an obscure solution to a global problem? *Plants, People, Planet* 3: 302–307.

Thorogood, C.J., Rumsey, F.J., Harris, S.A. & Hiscock, S.J. 2008. Host-driven divergence in the parasitic plant *Orobanche minor* Sm. (Orobanchaceae). *Molecular Ecology* 17: 4289–4303.

Thorogood, C.J., Rumsey, F.J., Harris, S.A. & Hiscock, S.J. 2009a. Gene flow between alien and native races of the holoparasitic angiosperm *Orobanche minor* (Orobanchaceae). *Plant Systematics and Evolution* 282: 31–42.

Thorogood, C.J., Rumsey, F.J. & Hiscock, S.J. 2009b. Host-specific races in the holoparasitic angiosperm *Orobanche minor*: implications for speciation in parasitic plants. *Annals of Botany* 103: 1005–1014.

Tóth, P., Undas, A.K., Verstappen, F. & Bouwmeester, H. 2016. Floral volatiles in parasitic plants of the Orobanchaceae: ecological and taxonomic implications. *Frontiers in Plant Science* 7: 312.

Turner, R.C. & Midgley, J.J. 2016. Sunbird-pollination in the geoflorous species *Hyobanche sanguinea* (Orobanchaceae) and *Lachenalia luteola* (Hyacinthaceae). *South African Journal of Botany* 102: 186–189.

Tzvelev, N.N. 2015. O rode zarazikha (*Orobanche* L. *sensu lato*, Orobanchaceae) v Rossii [On the genus *Orobanche* L. *sensu lato* (Orobanchaceae) in Russia]. *Novosti Sistematiki Vysshikh Rastenii* 46: 189–215.

Uhlich, H., Kreutz, C.A.J. & Rätzel, S. 2015. A contribution to the taxonomy and phytogeography of *Orobanche alba* Steph. ex Willd. (Orobanchaceae). *Phytotaxa* 222 (1): 1–16.

van der Ent, A. & Wong, K.M. 2015. Range extension of *Christisonia scortechinii* from mainland Southeast Asia into Borneo, and notes on the distinction between *Aeginetia* and *Christisonia* (Orobanchaceae). *Botanical Studies* 56: 28.

Wolfe, A. 2013. *Hyobanche thinophila* (Orobanchaceae), a new species from the Western Cape of South Africa. *Phytotaxa* 85: 56–60.

Wolfe, A.D., Randle, C.P., Liu, L. & Steiner, K.E. 2005. Phylogeny and biogeography of Orobanchaceae. *Folia Geobotanica* 40: 115–134.

Xu, Y., Zhang, J., Ma, C., Lei, Y., Shen, G., Jin, J., Eaton, D.A.R. & Wu, J. 2022. Comparative genomics of orobanchaceous species with different parasitic lifestyles reveals the origin and stepwise evolution of plant parasitism. *Molecular Plant* 15 (8): 1384–1399.

Yang, Z., et al. 2016. Horizontal gene transfer is more frequent with increased heterotrophy and contributes to parasite adaptation. *Proceedings of the National Academy of Sciences of the United States of America* 113 (45): E7010–E7019.

Yatskievych, G. & Contreras, L. 2009. A new genus of holoparasitic Orobanchaceae from Mexico. *Novon* 19 (2): 266–276.

Young, N.D., Steiner, K.E. & dePamphilis, C.W. 1999. The evolution of parasitism in Scrophulariaceae/Orobanchaceae: plastid gene sequences refute an evolutionary transition series. *Annals of the Missouri Botanical Garden* 86: 876–893.

Zázvorka, J. 2000. Orobanchaceae Vent. – zárazovité. In: Slavík, B., Chrtek, J. & Štěpánková, J. (Eds.), *Květena České Republiky* [Flora of the Czech Republic] 6: 477–513. Academia, Praha.

Zázvorka, J., Sánchez Pedraja, Ó., Moreno Moral, G., Carlón, L., Domina, G., Laínz, M. & Piwowarczyk, R. 2019. *Orobanche centaurina* Bertol. the correct name for *O. kochii* F.W. Schultz (Orobanchaceae). *Flora Montiberica* 75: 52–56.

Zhang, Z.-Y. & Tzvelev, N.N. 1998. Orobanchaceae. In: Wu, Z.Y. & Raven, P.H. (Eds.), *Flora of China*, vol. 18, pp. 229–243. Missouri Botanical Garden Press and Science Press, St Louis and Beijing.

INDEX OF BROOMRAPES

Accepted names are shown in *italic text* and synonyms in plain text. The main entry for each taxon is indicated by pink text. Images are highlighted by page numbers in **bold text**.

Phelipanche caesia
(left) photo: František Lamla

INDEX OF HOST PLANTS

Christisonia tubulosa
India (left) photo: Ambady Sasi